Reuse of foundations for urban sites

A best practice handbook

This Handbook is dedicated to the memory of Peter Starzec, our friend and colleague, who died tragically in 2006

Reuse of foundations for urban sites

A best practice handbook

Edited by

A P Butcher, J J M Powell and H D Skinner

Contributing authors

Tim Chapman, Sara Anderson

Ernst Niederleithinger, Alexander Taffe

Rab Fernie, Paul Tester

Gerard Evers

Aris Stamatopoulos

Göran Holm, Jenny Norrman, Wilhelm Rankka, Peter Starzec

Hendrick Ramm

Details of all publications from IHS BRE Press are available from:
www.ihsbrepress.com
or
IHS BRE Press
Willoughby Road
Bracknell RG12 8FB
Tel: 01344 328038
Fax: 01344 328005
Email: brepress@ihsatp.com

Published by IHS BRE Press

EP75

First published 2006
ISBN 1-86081-938-9 (10-digit)
ISBN 978-1-86081-938-4 (13-digit)

Foreword

In these days of heightened awareness of the impact we have on our environment and the way in which we manage it, we must not forget that we have to create the foundations for the future. In the past, we have been able to assume that we can largely ignore or remove what we have put in the ground previously. Now, with our ever-improving ability to construct large buildings on poor ground, particularly through the past 50 years, we have to consider the impact of what we have previously put in the ground on the cost and actual behaviour of new foundation systems. In addition, we must consider what impact these new foundation systems may have on future development. It is a matter that cannot be ignored.

As any archaeologist will tell you, the reuse of foundations is not a new subject and the principle of building on something that has already stood the test of time is a good one, provided that you understand the limitations of what you already have in the ground. In the past, a process of trial and error may have been acceptable under such circumstances. However, now, from Funder to Designer, we are risk-averse when it comes to foundations. We need to know what we are using to support our buildings, or at least be assured that, if we do not know precisely, we reduce the risk of anything happening to a level that we feel comfortable with. This requires the development of a suitable strategy to address the issues and give reassurance to all those involved in a project.

In London, we have since the late 1950s, been installing high capacity bored cast-in-situ concrete piles, deep into the London Clay, many of which were designed individually to take column loads. Many had under-reamed bases, some of which were dug by hand, some to the extent that they were touching each other. Such foundations effectively sterilise large parts of a site and may have a major impact on the scheming of a new development over the same footprint. Conversely, it makes no sense to avoid using them as they have proven high capacity. The answer must be to find ways of incorporating them within the new substructure. This requires some ingenuity and presents an exciting challenge to both geotechnical and structural engineers.

This is a new challenge. We have a range of tools at our disposal which can assist us in defining what is actually in the ground and how the new sub-structure may work. This *Handbook* is the product of a large collaborative project involving many professionals from a wide range of backgrounds. Much thought has gone into it! Its objective is to help all those involved in reusing foundations to understand the issues that need to be considered and to give some guidance and encouragement to all.

And finally, I hope that we all rise to the challenge, but whatever we do, that we learn from the experience and pass the information to future generations (preferably in the form of detailed electronic records that everyone can access when they need to!).

Hugh St John
Geotechnical Consulting Group
London

Acknowledgements

The RuFUS Project

The Reuse of Foundations for Urban Sites (RuFUS) research project was developed in response to a growing need to reuse foundations against a background of potential unknowns. It enabled research into perceived barriers, into a wide range of technical issues and facilitated sharing of experience in the reuse of foundations across Europe.

The European Commission Fifth Framework research programme, Energy, Environment and Sustainable Development, provided part funding. The project was funded 50% by the EC, the remaining funding being sourced by the partners.

Non-technical issues included case study evidence of successes and a robust risk and decision modelling approach. Clients and other professionals required a greater understanding of the issues in order to provide advice on insurance, archaeology and legal aspects.

The project aimed to provide 5 innovative technical developments for construction in inner cities, namely:

- measurement and analysis for testing of existing foundations beneath buildings to assess durability, integrity and geometrical shape,
- foundation loading performance of reused foundations,
- remediation/upgrading of existing foundations,
- 'smart' foundations for new foundations, and
- 'as-built' documentation system to future proof new foundations.

This 'best practice handbook' arises from the analysis of existing knowledge and areas of development.

Project Co-ordinator
Building Research Establishment Ltd (BRE), UK
www.bre.co.uk

Partners
Ove Arup and Partners Ltd (ARUP), UK
www.arup.com

Bundesanstalt für Materialforschung und – prüfung (BAM), Germany
www.bam.de

Cementation Foundations Skanska Ltd (CFS), UK
www.skanska.co.uk

Stamatopoulos and Associates (SAA), Greece
E-mail: secretary@saa-geotech.gr

Swedish Geotechnical Institute (SGI), Sweden
www.swedgeo.se

Soletanche-Bachy France
www.soletanche-bachy.com

Technische Universität Darmstadt, Germany
www.tu-darmstadt.de

BRE is grateful for additional financial support from the DTI and BRE Trust. The Project Partners are grateful for the support they have received from their sponsors.

EC Contract No: EVK4-2002-00099

Contents

1 Introduction

1.1 Foundation reuse is not new

Reusing foundations used to be the norm rather than the exception. Large structures whose siting was important, such as castles, tended to be rebuilt on the foundations of their predecessors. In Elizabethan times in London, in an attempt to curb urban sprawl, new building was only allowed if it was raised 'on old foundations' and later this was a common occurrence after the 'Great fire' of 1666.

As buildings have become bigger and expectations of their performance have increased, building occupants' acceptance of damage in structures has decreased. Structures themselves have also become less tolerant to differential settlements. Methods for calculating foundation requirements have become more reliable. All these factors have resulted in installation of new foundations for each new building to avoid aesthetic and structural damage caused by settlement.

Reuse of foundations can take many guises and does not always mean constructing a new building on old foundations. A common form of foundation reuse has occurred where the façades of a building are kept (for conservation or architectural reasons) and the internal parts of the building rebuilt (Figure 1.1). In these cases, modern construction components may allow more storeys to be included without an increase in load.

Recently, piles have been re-engineered and successfully reused on infrastructure projects, for example railway bridges and several major building projects (Chapman et al 2006), and several case studies (see *Section 1.3*, those included in *Appendix A* and Butcher et al 2006).

Figure 1.1 Façade retention for new apartment development

At the start of the RuFUS project in 2003, a questionnaire study assessed the level of awareness and understanding of reuse. Some 84 respondents from around the EU indicated that reuse was a relevant issue. Potential cost- and time-savings through reuse of existing foundations and avoidance of obstructions and archaeology were seen as opportunities, but technical and insurance issues were perceived as difficulties. Information on the old foundations together with investigation, assessment and design were seen as key technical areas where the RuFUS project could help reduce risks.

1.2 Sound engineering principles for foundation reuse

Foundations for any structure must be reliable, as demonstrated by an adequate factor of safety against failure. For a foundation system that has already been tested and 'proved' by the application of the first building load, a lower factor of safety against failure may be acceptable compared to that for new foundations, provided that sufficient details are known.

Foundations are designed to limit settlements. The settlement performance of a foundation system must be acceptable at working load, providing a factor of safety against damage in the structure that might impair its appearance or operation.

Existing foundations that are to be reused should be adequate for their intended purpose in the new building. The design of the foundations needs to be sufficiently robust so that it is no more likely to cause problems than installing new foundations. The requirements for reused foundations are no different from new foundations, and must be investigated, designed and incorporated into the construction so that these requirements can be met. Where foundation performance is critical (perhaps where large capacity is anticipated from reused foundations or where compatibility is required between old and new foundations), the observational method can be adopted to ensure robustness of design and construction. Verification of performance during and after

1.3 Case study on foundation reuse: Battersea Power Station regeneration

Client/Owner:	Parkview International (London) Plc
Architect:	Grimshaws
Structural & Geotechnical Engineer:	Buro Happold
Construction Management:	Bovis Lend Lease
Project Value:	£200m (Power Station Element)
Design Status:	Equivalent to RIBA Stage C (2006)

Overview

Proposals for the landmark Power Station at Battersea will offer an epic, vibrant space for new leisure, cultural, entertainment and retail facilities within central London. The Grade II listed building will also be complemented externally by new hotel, office and residential land uses as well as an extensive 3–4 storey underground car park. The option to reuse foundations within the Power Station has been undertaken by Buro Happold to deliver significant cost and programme benefits.

Courtesy of Parkview International (London) Plc

Ground conditions

The geology is typical of a Thameside central London site with drift underlain by a sequence of London Clay Formation, Lambeth Group and Thanet Sands overlying Upper Chalk around 70 m below ground. What makes the Power Station site unique is the extensive series of interconnected drift-filled hollows that penetrate up to 30 m within the London Clay Formation. The hollows, which are believed to be a glacial feature, are purported to be the deepest in London. Their complex geometry and variable infill has led to a series of foundations with unique characteristics that control their engineering behaviour.

Courtesy of Parkview International (London) Plc

Original construction

The Power Station was built in two phases because of the outbreak of World War II. The initial foundation construction (1930) was based on vibro piles a few metres into the London Clay Formation. However, the foundation is complicated by the fact that this was not always achieved and piles were also terminated within the overlying drift infill. The post war (1952) phase generally adopted large deep pad footings founded within either London Clay Formation or the overlying drift infill, apparently treated by a chemical injection process. The Power Station was decommissioned in 1983 and has already been subject to one abortive phase of redevelopment that modified a number of the existing foundations. 3D CAD modelling has provided a useful tool in establishing an accurate layout of the below-ground structure.

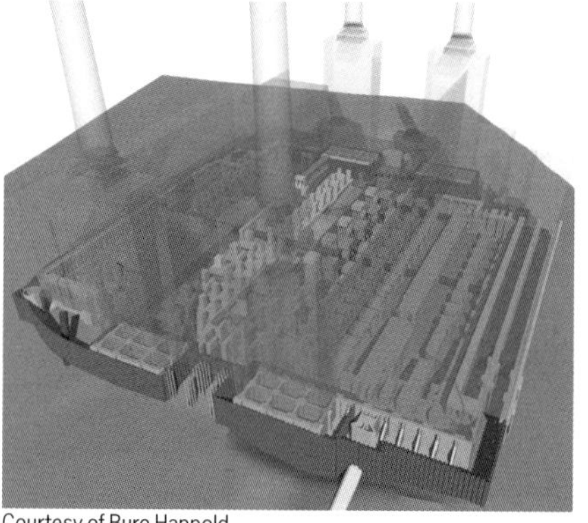
Courtesy of Buro Happold

Reuse strategy

Foundation reuse has been a core component of the design process from an early stage and has generally followed simple flow chart models to ensure that issues such as foundation integrity, durability, geometry and capacity are appropriately assessed at the key design stages. Considering the complex nature and variation in the foundations that control the differential settlement, Buro Happold had the strategy reviewed by the Project Geotechnical Advisor, Professor Chris Clayton, Southampton University. Preliminary discussions have also been held with Wandsworth Borough Council and Insurers to ensure compliance with regulatory and Insurance objectives.

Courtesy of Buro Happold

Investigation

Two significant phases of ground investigation have been undertaken to date.

The first phase was primarily undertaken to validate the foundation geometry as well as to establish the general integrity of the existing pad and pile foundations. The investigation included an electromagnetic survey, parallel seismic boreholes and pile integrity testing. A number of boreholes were also undertaken to refine existing assumptions about the geometry and characteristics of the drift infill. Generally, the geophysical methods proved useful in determining pile length but in some cases uncertainties remain.

The second phase of investigation was more focused on providing high quality soil parameters, particularly in relation to soil stiffness within the London Clay Formation and overlying drift infill. The follow-on work included self-boring pressuremeter, electric and seismic cone penetration testing, and small strain triaxial and bender element testing.

Courtesy of Buro Happold

Assessment

Analysis has needed to recognise the challenge of matching stiffness to limit deflections within new structures, particularly with mixes of pile groups and large foundations. Simple VDSP modelling has been undertaken to examine global settlement predictions and to highlight key risk areas. Areas of critical movement have also been subject to 2D and 3D finite element (FE) modelling using PLAXIS. The results have been fed into the structural analysis on an iterative basis.

Courtesy of Buro Happold

The next stage

Residual uncertainties identified from ground investigation are generally better addressed through the final design stages and construction management process. Proposals are also in place to undertake a final phase of investigation before construction, including pile load testing at carefully selected locations. The pile testing will validate historic load test results as well as examine differences in performance of foundations installed within the London Clay Formation or drift infill.

Pile testing will be undertaken in parallel with final design to maximise foundation reuse, albeit it is expected that a limited number of new bored piles will be required to accommodate the intended structural grid. The testing will also be supported by structural monitoring to keep movements within acceptable limits. The results of monitoring will also be fed into the Construction Management Action Plan to ensure that if unexpected results are encountered, a clear strategy is in place to deliver timely remedial measures, should they prove necessary.

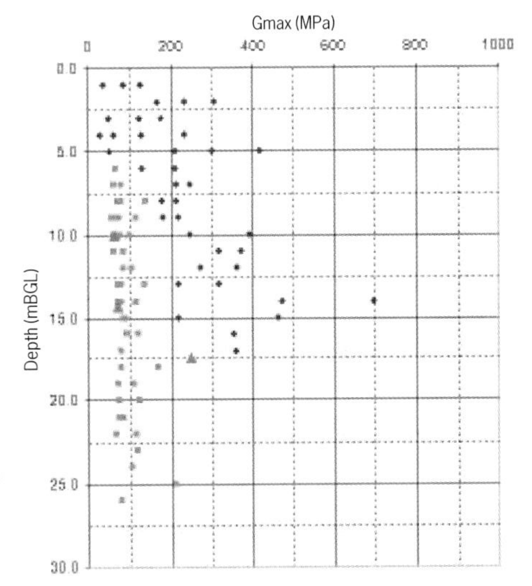

Courtesy of Buro Happold

Further details can be obtained from:

Justin Phillips
Associate Director, Buro Happold
Camden Mill, Lower Bristol Road
Bath BA2 3DQ

Tel: 01225 320600
Fax: 0870 787 4148
Email: justin.phillips@burohappold.com

construction is carried out against defined criteria; remedial actions are determined should the criteria be exceeded.

The reuse process should be carried out in parallel with other design and investigation decisions and activities. This will maximise the opportunity for effective reuse and its advantages while minimising the risk of increased development time should reuse prove not to be the most appropriate option.

1.4 Why foundation reuse can be a concern

Important technical issues must be addressed to ensure that foundation reuse is undertaken appropriately. However, there can be perceptual and practical reasons why foundation reuse may cause concerns that do not arise with a new foundation system: for example, some parties involved in a development may have the (completely unfounded) perception that reusing foundations is in some way 'second best' and that 'newer is better'.

Foundation reuse is not explicitly addressed in current foundation design codes such as BS 8004 or BS EN 1997-1. Foundation design using reused foundations often may not fully comply with current codes, particularly for materials, so a pragmatic approach is required. The client may have to allow a dispensation to the design team in respect of certain elements not being built to current codes.

The amount of investigation of a reused foundation system may need to be balanced against perceived risks as well as the amount and veracity of information. The design of foundations that are reused or that incorporate reused elements may require explicit assessment of settlement, to address the uncertainty inherent in current understanding of foundation behaviour.

The use of foundations that already exists clearly places some constraints on the building designers (both foundation and structure) and this may cause concern.

Apart from technical issues, there are issues relating to the provision of insurances and warranties that need to be addressed.

1.5 How to use this Handbook

This *Handbook* provides technical guidance on the processes for deciding on the correct foundation options, and for investigation, design and construction using reused foundations. It aims to give the reader a sound understanding of the background and key issues in foundation reuse and how to address risk through sound engineering and decision making.

Foundation reuse is a specialist area of geotechnics and appropriate advice should be sought at an early stage of a project to minimise risks.

Box 1.1 is intended to guide the reader to the relevant chapters in the *Handbook*, related to the stage of development of a project and the information necessary at each stage.

Box 1.1 How to use this Handbook

Chapter	Project stage: Planning	Design	Construction	Verification	Completion	Further reading and detailed examples
Chapter 2: Drivers for reuse	Why reuse foundations? Advantages and opportunities					Case histories, Appendix A
Chapter 3: Key technical risks	← What to beware of and how to reduce risks →					Economics and risk, Appendix B and C
Chapter 4: Legal and financial context	← How does reuse sit within the current systems for projects? →					Idealised case demonstrating risk modelling, Appendix B
Chapter 5: Decision model	← Should you reuse? → What to consider and how?					
Chapter 6: Investigation, assessment and design of reused foundations		← How to assess, investigate and design to reuse →				Flow charts, Appendix D Foundation performance and its improvement, see Butcher et al 2006
Chapter 7: Design of new foundations for future reuse	← How to ensure foundations can be reused next time →					Typical documentation, Appendix F Smart instrumentation, Appendix E

1.6 References

British Standards Institution. BS 8004: 1986 *Code of practice for foundations*

British Standards Institution. BS EN 1997-1: 2004 *Eurocode 7. Geotechnical design. General rules*

Butcher AP, Powell JJM & Skinner HD (eds). *Reuse of Foundations for Urban Sites: Proceedings of International Conference*, BRE, Watford, 19–20 October 2006. Bracknell, IHS BRE Press, 2006

Chapman T, Anderson S & Windle J. *Reuse of foundations.* CIRIA Report CP/107. London, CIRIA, 2006

2 Drivers for reuse

2.1 Introduction

This chapter outlines why developers, building owners and construction professionals should consider reusing foundations. In the past, the reuse of foundations has only been considered if congestion in the ground rendered other solutions impractical. However, changes to legislation, improvements in our technical understanding, recognition of the needs of the environment, and sustainability are driving us to reuse foundations. These influences are becoming more powerful and can even be used to a developer's advantage by:

- enabling cost-effective redevelopment,
- allowing higher rents to be charged for occupying 'environmentally friendly' buildings, or
- enabling the process for gaining planning consent.

The various current and future drivers that influence the reuse of foundations on urban sites include:

- ground congestion,
- requirements to preserve archaeology,
- technical drivers,
- economic factors, such as energy and natural resource costs,
- changing economics of demolition and construction,
- environmental drivers supported by legislation,
- promotion of sustainability.

In order to make judgements about the drivers for the reuse of foundations the reader is urged to use the decision models described in *Chapter 5*.

2.2 Ground congestion

Ground congestion is and has been one of the prime drivers for the reuse of foundations in urban areas and will become more critical as urban centres become ever-more developed (Figure 2.1). The working life of buildings in the financial centres of cities has been reducing so that at present their anticipated life is often substantially less than their nominal life of 50 years, and is sometimes as little as 20 years. With time, the facilities provided in the building may not meet the latest expectations or needs of occupants or building

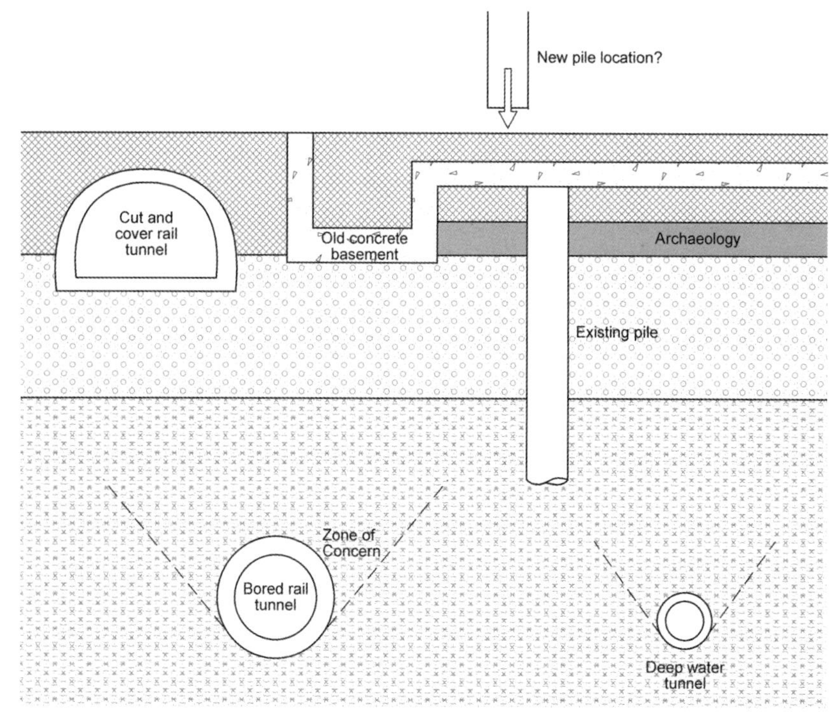

Figure 2.1 Diagrammatic representation of congested ground

regulations. At this point it may no longer be viable to refurbish a building so redevelopment may be the most economic option. The ever-more rapid redevelopment cycle results in more ground congestion driving the reuse of deep foundations.

2.2.1 Old foundations

Physically removing old piled foundations to make way for new foundations can cost significantly more than (up to 4 or more times) the cost of constructing new piles. To design an effective method for removal, details of the geometry of any pile to be removed will have to be determined. Some piles may be difficult, if not impossible, to remove. The removal of old piles disturbs the ground and the voids created have to be stabilised by backfilling with material that will allow both the new foundation to be constructed and the surrounding ground to be supported. This can lower the capacity of the new piles.

If the old pile locations do not fit with the load points of the new structure, transfer structures can be designed to enable existing piles to be incorporated into the new structure. The capacity of old piles can be enhanced, if required, to facilitate their use in a new foundation arrangement (see *Chapter 6*).

2.2.2 Infrastructure and service tunnels

Infrastructure (road, rail and pedestrian) and service (water, sewer, power) tunnels are extremely expensive to move, repair or replace so their existence will force any new development to avoid locating foundations in positions that could adversely affect them. In all cases, the owners and operators will have to be notified about the new development and, in particular, they will want evidence that the change in loading and induced settlements on their tunnels do not cause unacceptable effects. They are likely to have statutory powers to restrict the construction of new foundations unless it can be demonstrated that they will not affect the safe working of their tunnels. Significant amounts of work, including extra ground investigation, modelling of the design and monitoring during and after construction will be required to satisfy the tunnel owners/operators that the new construction will be acceptable. In many cases, the use of pre-existing foundations can ease these development problems near infrastructure.

2.2.3 Underground services

Smaller, shallow service tunnels that were constructed by a trench method can be moved but at a cost. In some cases, such as fibreoptic communication cables, these costs can be high or in the case of a gas main there will be periods of high gas usage in the middle of winter when alteration is plainly impossible. In addition to construction costs, additional costs associated with the interruption of supply and the risk of damage need to be included in the project cost and risk assessments.

2.3 Archaeology

There is a continued desire to preserve archaeological artefacts that are particularly valuable in the understanding of historical events. All urban centres are likely to have an archaeological archive beneath the surface of the ground. In many cases, knowledge exists of the whereabouts and quality of the archaeology but in other cases, there is no existing knowledge. In many urban centres, local byelaws ensure that excavation of the ground is notified to the appropriate authorities in order to protect archaeology. In some particularly important archaeological areas, the type of foundation allowed to be installed is restricted to those that inflict the least damage in terms of archaeology destroyed or displaced.

Archaeology and other cultural heritage elements such as standing buildings may be damaged by piling operations and other foundation construction techniques. Increasingly, throughout Europe, cultural heritage will be considered through the process of strategic environmental assessment, and environmental impact assessments should identify the presence of archaeology in advance of any development project.

In different EU countries, policies differ as to how to deal with archaeology when it is encountered on construction projects. In part, this is a reflection of the type of archaeology. For example, in Mediterranean countries, such as Greece, often substantial evidence consisting of stone buildings and other structures can be found, representing many millennia of human development and civilisation. By contrast, the buried archaeology of northern Europe can be more ephemeral, with buildings often constructed using less durable materials.

In Greece, archaeological sites are often preserved in situ by incorporating the remains of the site within the new development. In other instances, piling through areas of lesser archaeological interest allows the greater proportion of the site to be preserved (Stamatopoulos et al 2006). In that case, only 2% of the site was lost through piling. This scenario is also common in the UK), where planning policies covering archaeology recommend the in-situ preservation of nationally significant remains (Department for Communities and Local Government 1990). There are a limited number of examples of structural elements (walls, floors, etc.) that have been preserved and displayed within buildings (eg the remains of London's Roman amphitheatre preserved and displayed beneath the new Guildhall Art Gallery (Ganairis & Bateman 2004), since these types of remains are rare. More commonly, archaeology is preserved in situ below ground and is not visible.

There are two main topics under which cultural heritage should be considered in relation to foundation reuse. First, foundation construction damages archaeology (Biddle 1994). Therefore, foundation reuse in situations where buildings have previously been constructed on archaeological sites will reduce future damage from new foundations. On such sites, a feasibility study to look at foundation reuse should be carried out as part of any re-development programme. As a subsidiary issue, in towns and cities with fragile historic buildings, some foundation installation techniques may risk causing vibration damage to these structures. In these situations, foundation reuse would also reduce this risk.

2.4 Technical drivers

The technical drivers are those where the advancement of understanding the engineering behaviour will allow higher loads to be supported by an existing foundation without compromising risk. In some cases, a better understanding of the previous structure may reveal that a lower design load will be imposed by the new structure. Other changes may impact new designs, for example, the requirements for groundwater control or proximity of other buildings and the need to limit ground movements.

2.4.1 Full use of existing design capacity

The existing design capacity of a foundation can be fully used if the original design philosophy and actual imposed loads can be determined. An existing foundation can have a reserve of capacity, particularly where overcapacity was built into the original design as an enhanced factor of safety to account for a lack of knowledge or understanding.

The existing foundation design would usually have been completed using the understanding of structural behaviour, soil–structure interaction and foundation behaviour at the time of the design. However, continued research and experience extends and deepens understanding, and can lead to the critical design criteria changing. For example, the design of deep excavations in heavily overconsolidated clay soils now uses small strain behaviour of soil and numerical analysis to model behaviour, and realistic local deformations result from the design predictions. Older design procedures may not have used advanced modelling and been calibrated by concurrent monitoring, and so were more likely to produce a structure that was significantly over-designed. In the past, certain components of pile capacity were often omitted from the design, such as the shaft friction of an under-reamed pile, because the base load was sufficient and full understanding of the load take-up of the shaft was not appreciated. In other cases, the vertical load capacity of a retaining structure may have been ignored because the interaction of the foundations and the retaining structure was not understood at the time of the design.

The stress–strain behaviour of soil is typically non-linear and so designers prefer to use only a small percentage of the available capacity to control settlements. The loads used in the design of a foundation would have been calculated by the structural engineer who would have included factors of safety so as not to over-stress the structure or the foundation. The foundation engineer would also have included a factor of safety so that a typical foundation might only be working at 20% capacity or less. If the settlements can be controlled there will, in general, be significant foundation capacity that has not been used in the existing building that could be harnessed for the new structure.

2.4.2 Increased capacity of old foundations

Research has found that piles installed and tested between 20 and 30 years ago have, upon retest, gained in capacity with a significantly stiffer initial loading behaviour (see *Chapter 6*), ie they show less settlement for the same loading. Other piles tested after being loaded to working loads for the life of a building and then tested after demolition of the building have shown significantly stiffer behaviour and higher ultimate loads (see Butcher et al 2006). This increased capacity and stiffness could be used in the reused foundations because settlement control can be obtained at higher working loads than the previous building.

2.4.3 Potential for increasing the capacity of old foundations

The capacity or settlement performance of foundations can be enhanced to facilitate a change of use. In the UK, grouting works have been carried out to upgrade foundations and reduce settlements before nearby tunnelling is undertaken. A number of options for upgrading are described in this *Handbook*.

2.5 Economic factors

Economic drivers for the reuse of foundations occur throughout a redevelopment project. Box 2.1 lists both savings and potential extra costs that need to be considered.

At the start of the project the reuse of foundations has the potential to:

- reduce the cost of site investigation,
- cut project demolition time, and
- cut project construction time and costs.

Any reduction in demolition and construction time allows earlier occupation of the new building and thereby earlier rental income.

2.5.1 Reduced cost of investigation

All sites need an investigation to assess:

- what is in the ground,
- how it can be used to support the proposed structure, and
- the ground conditions themselves.

Box 2.1 Potential savings or costs in the reuse of foundations

Savings	Potential extra costs
• Ground investigation • New foundation design • New foundation construction • Spoil disposal • Raw and processed materials • New foundation construction time • Environmental impact of new foundation construction	• Investigation of existing foundations • Remediation of existing/additional foundations • Redesign/relocation of superstructure • Interfacing with new superstructure • Environmental impact on the operation of the new building

On a site with the potential to reuse the foundations the desk study stage of an investigation (see *Chapter 6*) becomes even more important than usual. In general, it is much less expensive to collect existing information than generate new information from physical activity. The collection of desk study information enables effective scoping, planning and timetabling of investigations as well as providing information for other stages of the construction and reuse process.

The traditional site investigation methodology is to test the soil, in situ and in the laboratory, and use the soil properties to design the foundations. On a site where there is the potential to reuse the foundations, the investigation needs to be carefully designed to assess the soil properties (particularly if supplementary foundations may be needed, see *Chapter 6*) but also the properties of the foundations to be reused. Depending on the type of information obtained from the desk study, the geometry and integrity of the existing foundations will need to be assessed. Savings may accrue from a reduction in the extent of investigation and/or non-intrusive techniques to assess the existing foundations may be significantly less expensive than traditional investigations. The actual costs of the various activities will vary from site to site depending on the size of the site, degree of reuse and level of information available before the site works.

2.5.2 Project cost and time savings

Construction works are an expensive part of a redevelopment scheme and foundations are on the critical path for a project, particularly in urban areas where sites are restricted and congested. Any activities that can be reduced, carried out concurrently or avoided will reduce the project time and costs. Figure 2.2 shows a diagrammatic representation of some of the major activities in foundation construction and an estimate of what their relative costs might be.

A significant portion of time spent on a redevelopment project will be on deconstructing the existing building, the ground floor or basement and probably the foundations. This is true particularly if the planned positions of the proposed new foundations coincide with the positions of existing ones.

The time to break out the top reinforcement in the existing foundations to tie in new works can be less than constructing a new pile.

Any reduction in the project time reduces the time the owner/developer is either paying interest on the capital outlay for the redevelopment or not receiving rental income.

To take a balanced view, any savings by reusing foundations may have to be offset against extra costs for transfer structures, the extra cost of demolition in preserving pile tops, and checking for defects in piles and reinforcing steel.

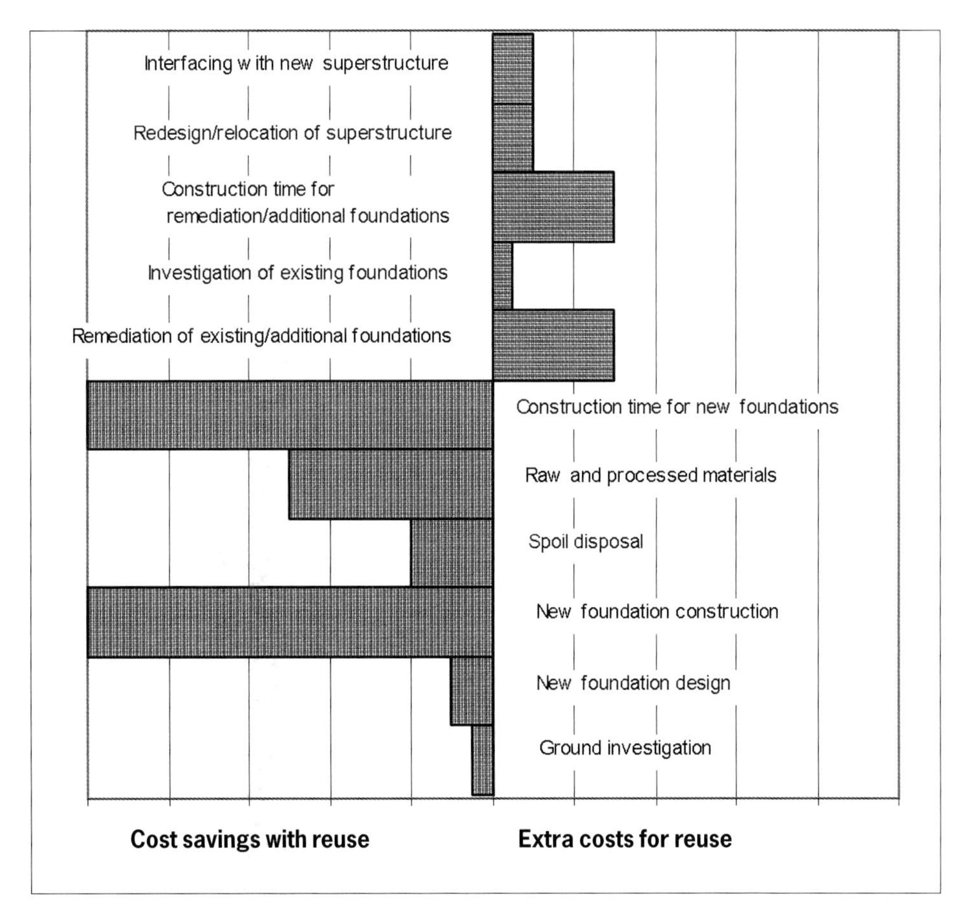

Figure 2.2 Diagrammatic representation of some of the major acitivites in foundation construction and projected costs

2.6 Changing economics of demolition and construction

The cost of disposal of waste materials produced from demolition is set to continue to rise through legislation of waste repositories that will restrict the type of waste to be accepted and the increasing pressure to recycle materials. Increasing energy costs will push up transport costs for the removal and supply of materials. The costs of producing construction elements will also increase with the increase in energy costs needed to source raw materials.

2.7 Environmental drivers

There are many issues to be resolved surrounding construction and the environment with increasing pressure to minimise the impact of construction. The reuse of foundations can significantly reduce the impact of the redevelopment of urban areas on the environment.

The reuse of foundations can:

- reduce the use of natural resources,
- reduce the total energy used,
- reduce the potential for groundwater level changes and groundwater pollution during construction,
- reduce the quantity of waste materials produced both during demolition and construction,
- reduce the number of vehicle and plant movements.

The reuse of foundations will cut the use of natural resources simply because new foundations will not be made. The natural resources used in the production of steel and concrete are generally non-renewable, though the reuse of aggregates and steel is growing. However, the production processes for producing steel and concrete use other natural resources not required in the recycling of materials.

The total energy used on a reuse of foundations site will be reduced since energy to manufacture, transport and place the constituent materials for the foundations will not be required. Further, if existing foundations were to be removed the potential energy use in their removal and disposal of the materials will be avoided.

Vehicle and plant operations use energy (thus consuming non-renewable natural resources) and contribute to air and noise pollution. The redevelopment of urban centres will, by their nature, be in sensitive areas for air and noise pollution and may be subject to local restrictions on working practices and times.

2.8 Future sustainability

To incorporate into the design of any development scheme, whether redevelopment or new, features that will enable reuse of foundations (perhaps more than once) in the future will contribute to sustainable construction. Developers can safeguard the future redevelopment potential of their land (and its value) and can ensure that urban centres can be maintained.

The increasing costs of energy will require buildings to be more efficient and have a significantly longer life. The design of new buildings will be forced to be versatile to meet the changing needs of occupiers and to provide sustainable accommodation.

2.9 Summary

The reuse of foundations can:

- avoid or reduce the problems of ground congestion,
- generate economic advantages,
- preserve archaeology,
- reduce environmental impact of the new building,
- be achieved through greater technical understanding of foundation and structure behaviour,
- promote sustainability.

2.10 References

Biddle M. What future for British archaeology? Opening address, *Archaeology in Britain Conference, 8th Annual Conference of the Institute of Field Archaeologists*, Bradford, 13–15 April 1994. Oxford, Oxbow Books, 1994

Butcher AP, Skinner HD & Powell JJM. Stonebridge Park: a demolition case study. In: Butcher AP, Powell JJM & Skinner HD (eds). *Reuse of Foundations for Urban Sites: Proceedings of International Conference*, BRE, Watford, 19–20 October 2006. EP73. Bracknell, IHS BRE Press, 2006

Department for Communities and Local Government (DCLG). *Planning Policy Guidance Note 16: Archaeology and Planning*. London, The Stationery Office, 1990

Ganairis H & Bateman N. From arena to art gallery: the preservation of London's Roman Amphitheatre in situ. In: Nixon T (ed) *Preserving archaeological remains in situ? Proceedings of the 2nd Conference*, Museum of London, 12–14 September 2001. pp 198–201

Stamatopoulos AC, Stamatopoulos CA & Photiadis MG. The new Acropolis Museum of Athens. In: Butcher AP, Powell JJM & Skinner HD (eds). *Reuse of Foundations for Urban Sites: Proceedings of International Conference*, BRE, Watford, 19–20 October 2006. EP 73. Bracknell, IHS BRE Press, 2006

3 Key technical risks

3.1 Introduction

There are risks inherent in the reuse of foundations, but these can be controlled by appropriate risk assessment and careful design. The design process requires a good understanding of the geotechnical and structural considerations of the foundations together with an appreciation of the behaviour of the structure above. The extent of analysis needed depends on the particular risks. Where good information is available on the extent and condition of the old foundations and the amount of reload is within the assessed working capacity, the risks are lower. However, where the risks are judged to be higher, greater care is needed during the process.

3.2 Structural damage in buildings due to inadequate foundations

If foundations fail, whether new or reused, the structure will undergo unacceptable movement leading to damage. Damage may be minor, perhaps superficial or decorative cracking, or may be more substantial and lead to partial or complete collapse of the structure.

Excessive foundation and structural movements relative to the surrounding ground can result in impairment or failure of services running from the surrounding ground into the structure, cracking or other deformation of the pavement around the building and uneven step heights for steps accessing the building.

Differential settlement between adjacent foundations may result in cracking of the structure or its façade. Brittle façade materials such as glazing are particularly susceptible to damage. Differential settlements may also affect the functioning of the building such as doors jamming, or piped services being impaired through substandard falls.

Excessive deformation of part or all of the structure could lead to collapse of the structure. Collapse of the structure is not only dependent on the failure of particular foundations but will also be affected by the robustness of the overall structural design. Buildings that are unable to distribute any loading between different columns or buildings with a single foundation for each column will be more susceptible to the poor performance or failure of a single foundation than a more robust building.

Foundation reuse is unlikely to change the manner in which foundations fail, but for less well considered reuse of foundations, the likelihood of damage may increase. The risks associated with this increased likelihood of deficient foundation performance should be managed and reduced as part of an overall risk management process as identified in *Chapter 3* and using the design approaches outlined in this chapter and *Chapter 5*.

Methods of reducing the risks associated with foundation reuse include:

- review of all available information on the foundations, and
- investigation of existing foundations.

Mitigation methods include:

- design of a robust superstructure to reduce the impacts of a single pile failure,
- adoption of an appropriate level of caution applied to the reuse load-carrying capacity,
- design of a partial reuse scheme with a robust foundation design.

3.2.1 Types of structural failure

It is important to recognise that different types of 'failure' can occur that must be recognised and addressed in the design process.

3.2.1.1 Limit state design

In their commentary on Eurocode 7, Simpson & Driscoll (1998) describe limit state design as:

> 'a procedure in which attention is concentrated on avoidance of limit states, ie states beyond which the structure no longer satisfies the design performance requirements'.

This definition of limit state is essentially practical and relates to the possibility of damage, economic loss or unsafe situations. For example, cracking or distortion (which may have no more consequence than giving a disappointing appearance) constitutes a limit state, just as does a catastrophic failure.

Two limit states are adopted in most limit state design codes: serviceability and ultimate.

3.2.1.2 Serviceability limit state (SLS)

Serviceability limit states correspond to conditions beyond which specific performance requirements, usually settlement-based, are no longer met. Typically, service requirements relate to allowable deformation to ensure that in normal operation doors open freely and do not jam, the building envelope and finishes do not crack thereby letting in water or looking unsightly, and drains run in the correct directions with suitable gradients. Some damage may be visually displeasing and could cause concern to the building occupants. The operation of the building is likely to be affected and therefore will be an annoyance for users.

The serviceability limit state may correspond to some form of damage but should occur with no risk to life.

The implications of damage associated with SLS failure may include:

- disruption due to impaired function of the building,
- disruption due to the building being evacuated until the damage can be clearly identified as being limited to a serviceability failure and not warning of an impending collapse,
- disruption during repairs of the building.

When damage occurs, there will often be a dispute about responsibility and who should pay for the costs associated with disruption. If lawyers and experts are included, the costs associated with the damage and dispute will increase further; in the past it has been common for these costs to outweigh the direct costs by up to a ratio of 5:1 in the UK (Chapman 2004).

3.2.1.3 Ultimate limit state (ULS)

Ultimate limit states are those associated with collapse or with other similar forms of structural failure. They are concerned with the safety of people and the safety of the structure. Structures must have an adequate margin against collapse for the range of possible loads that might be applied and the building must remain standing. While building damage may occur and therefore limiting deflections are less applicable, ULS may include excessive deflections leading to collapse. Examples of collapse include beams losing bearing on column heads or columns buckling.

Costs that may be associated with ULS failure are listed below.

- *Repair and rebuilding costs.* These are likely to be substantial as they will include demolition and construction costs.
- *Business disruption costs.* These are likely to be substantial, in particular if the building failure is rapid and the building cannot be evacuated of all business-related information. All data and computer storage could be lost, which could have a significant detrimental effect on business continuity. Customers will be lost until replacement facilities are available.
- *Compensation for those killed or injured.*
- *Dispute costs.* Often there will be a dispute about responsibility and who should pay for the other costs resulting from ULS failure. As part of that exercise, lawyers and experts may be appointed, which will inflate the overall cost — it has been common for these costs to far outweigh the direct costs.

3.2.1.4 Examples of failure

Types of foundation failure that may lead to a limit state being breached include those listed below.

- *Direct bearing capacity failure* resulting in excessive foundation settlement. The column above the foundation will also settle, resulting in beams sagging, floors tilting, services breaking and a damaged façade (SLS/ULS).
- *Single foundation element settles more than expected.* Some local damage may occur which is unsightly and worrying for occupants (SLS).
- *Part of a foundation tilts,* perhaps due to excessive settlement of one pile in a multiple-pile group. The column may tilt and cause beams to lose bearing (SLS/ULS).
- *Defects* (eg omission or deterioration of reinforcing steel) may lead to foundations being unable to resist horizontal loads (ULS may occur).

3.3 Stages for risk reduction

Chapter 5 identifies a decision process for a construction project considering the reuse of old foundation. Risk management is an integral part of this and is carried out through a series of stages as follows.

- *Identification of potential risks through:*
  - recognition of principal determinants of foundation reliability and reuse context,
  - appraisal of old building operation,
  - foundation reuse desk study of available information and knowledge,
  - investigation of old foundations.
- *Assessment of identified risks through:*
  - assessment of sufficiency of investigation,
  - foundation reuse design approach.
- *Control through:*
  - evaluation of structural strategy for the new building,
  - control during construction of foundation system for new building.
- *Monitoring:*
  - monitoring of new building to confirm design performance.

3.3.1 Foundation reliability

The reuse capacity of existing foundations will be inferred initially from the available data. For well-documented projects where the construction records, including drawings and non-conformance reports, are preserved, there will be a considerable degree of confidence not only in the physical presence and properties but also the constructed quality of the foundations.

The reliability of existing foundations may depend on the future requirements as well as on their present condition. For single piles, Figure 3.1 shows the features that may determine reliability for future use.

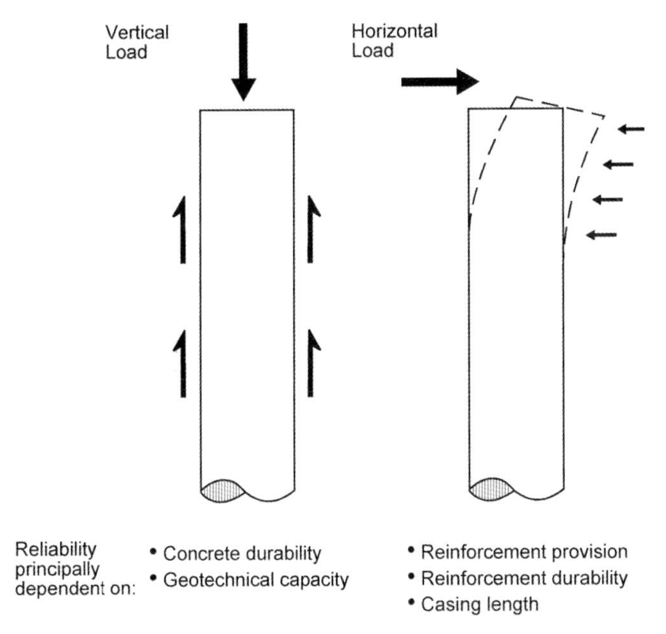

Figure 3.1 Reliability of existing foundations

The application of eccentric loading to the foundation system will induce tension and horizontal loading in the old foundations. The resistance of such loads is dependent on the provision of appropriate reinforcement and on its condition. This could be expected to vary across the foundation system, may not be adequately recorded on drawings and cannot easily be verified by site testing. Reuse of foundations to support tension loads, horizontal loads or applied moments can be more problematic and involve higher risk than compressive loads.

Where good records are unavailable or where only design and not construction information is available, potentially the available information could mislead the design team as to the possible properties and quality of the installed foundations.

Foundations not present

Design changes are not uncommon both immediately before and during construction. If the only records available are based on a design that was not constructed the information can incorrectly indicate the presence of particular foundations which may have been modified, moved or even eliminated from the design of the foundations.

An assessment of the previous structure and applied loading may give some indication of the foundation requirements but will be unable to indicate conclusively the presence and location of foundations.

In the most extreme case, the absence of an existing foundation to be reused as part of the foundation solution represents a major flaw in a design.

Major pre-existing flaw in foundation

Major pre-existing flaws may be present in the foundations. These flaws may have been introduced:

- *during construction of the foundations.* Where noted on site these may have been recorded in construction documentation such as non-conformance reports and the remedial steps may be known. Where there was limited supervision during installation, the flaw may not have been noticed and no remedial steps may have been taken. Without records, the confidence that construction flaws were found and resolved is reduced.
- *by damage after construction,* such as overloading of the foundations.
- *by poor design of the foundations,* such as the provision of insufficient structural or geotechnical capacity which will result in overloading under the original design loads and may also be associated with additional settlement of the foundations.

Good construction and performance records in conjunction with an assessment of the existing building's load settlement performance are the means of identifying potential major pre-existing flaws in foundations, before the investigation of the foundations.

The design process must address aspects of reliability such that the reused foundation scheme presents acceptable risks and typically should be as reliable as the use of a new set of foundations.

3.3.2 Appraisal of old building operation

It is important to confirm that the old building performed satisfactorily. The building should be checked for evidence of unusual or differential settlement (for instance by observation of the façade for evidence of repairs, by checking records for any remedial works, or by levelling a building course, eg at damp proof course level around the building perimeter). For internal surveys it should be borne in mind that buildings that have suffered from settlement may have had the floor refinished to provide a level surface and therefore a survey of the ceiling may provide a better indication of movement. Sometimes clients will have maintenance records for the whole life of the building, or there will even be an individual who remembers all repairs that occurred. For other buildings, records may be sparse or unreliable, so the redevelopment team will have to examine the building closely for signs of damage that could be due to movement of the foundations.

For buildings where foundation under-performance is suspected, and where the foundations are proposed for reuse, it is vital to diagnose correctly the cause of any previous structural damage to the first building. Existing building defects which are suspected as having been caused by a foundation failure should have a profound effect on the confidence of placing a new building on the existing foundations. As only a small proportion of reused foundations can generally be checked comprehensively, any observations of structural damage in a first building may impact on the reliability, and hence design, of the rest of the foundation system. A greater extent of testing than normal would be required to provide confidence before reusing the foundations from a building where the first set of foundations did not appear to have functioned adequately. Otherwise, only a smaller proportion of load can be safely applied.

A second check that is required is to assess the extent to which the building is likely to have been loaded over its first life. The history of loading of the building should be assessed to determine if it is currently fulfilling the same use for which it was originally built or whether, over its history, it is likely to have been loaded to a greater extent than presently. Consulting old trade directories may give useful information on previous building uses which will provide an indication of historical loading conditions. An example of this would be a warehouse which has been converted into a hotel, offices or a residential building.

If the live loading is likely to have been high, but may reduce in the next life, there is likely to be reliable excess capacity available for structural renovations. Sometimes a history of the building's usage can be compiled from archive sources like old trade directories or local newspaper archives.

In many cases, the appraisal of the old building operation will be incorporated within a foundation reuse desk study for the site as indicated in the foundation reuse decision process flowcharts in *Section 5.3.*

3.3.3 Foundation reuse desk study of available previous knowledge

A desk study is an integral part of any development design process and risk identification and management process. Where foundation reuse is an option, the desk study forms a particularly key stage in the risk identification process and therefore the importance of a thorough desk study cannot be overstated. In some countries, the process of undertaking a desk study is not standard practice for building projects, however, the requirements of Eurocode 7 (EN 1997-1: 2004), clause 3.1 (1)P effectively define a standard desk study:

'Careful collection, recording and interpretation of geotechnical information shall always be made. This information shall include geology, geomorphology, seismicity, hydrology and history of the site. Indications of the variability of the ground shall be taken into account'

In addition to more standard information such as site history, archaeological and contamination potential, a foundation reuse desk study should collate all of the available information on the existing building, including the foundation system, to understand how the structure works. Information collected should include as-built information (eg any drawings and non-conformances from the installation of the existing foundations). This will help manage the risk of poor construction quality of the existing piles which will increase confidence in foundation reuse for the project.

The information on the existing structure that should be sought as part of the foundation reuse desk study is listed in Box 3.1 and expanded in *Chapter 6.*

The foundation reuse desk study should specifically record:

- the status of the information found and whether it is comprehensive or likely to be a reliable indicator of the extent of what was constructed,
- any information which is specifically missing from the available information.

Box 3.1 Information on the old building to be collected in a foundation reuse desk study

Desk study and site investigation
Previous desk study reports
Factual ground investigation report
Geotechnical interpretative report
Test results
Records of water level monitoring
Archaeological reports

Design and specification
Design philosophy and codes
Geotechnical design
Structural design of pile elements
Force combinations applied to each pile
Construction specification

Construction
As-built pile location plan
As-built pile schedule, giving diameters and other pile details
Pile construction record sheets
Pile reinforcement schedule
Details of pile integrity testing
Results of pile load testing
Details of all non-conformances and how they were resolved
Confirmation from an independent body that the records are correct
Results of settlement monitoring
Construction and post-construction photographs

Building operation
Structural alterations
Maintenance records
Observations of damage

Demolition
Load take-down

The absence of some information may make it imperative that a more thorough investigation of the foundations should be carried out if the foundations are to be reliably reused.

A good foundation reuse desk study should attempt to locate all old construction records for the building. Sometimes it may be necessary to think laterally to find the correct place where old records have been saved. Possibilities include those listed below.

- *The Building Control offices of the Local Authority or other regulatory approval authority.* Sometimes, records are disposed of due to a lack of space and increasingly access to the records will require written authorisation from the building owner.
- *The original designers:* architect or structural engineer, although often the practices have merged or gone out of business. A fee will often be charged to cover the costs for locating and copying the documents. Sometimes, the fees may be set at a level commensurate with the perceived value of the records to the new developer. Often, the available records will be limited or so sparse as not to be helpful.
- *The original foundation contractor,* although similar restrictions may apply as in the previous point.

An example of good luck in locating old records during an investigation is given by St John & Chow (2001). They recount how in 2000 a time capsule, buried on 21 September 1961 when the building was being built, was found in a footing being broken out in a trial pit to appraise the extent of existing foundations. Among other artefacts in the capsule was a drawing showing the foundation layout. Few other developments will have similar good fortune.

3.3.4 Investigations of old foundations

Investigation of the existing foundations will be required. The scope and detail of the investigation will be dependent on:

- the information available on the existing foundations identified in the foundation reuse desk study process,
- the quality of the information,
- the expected condition of the pile population, and
- the reliance to be placed on the old foundation to support the new structure.

The findings of the foundation reuse desk study should be used to design the investigation and, in particular, to identify specific foundation locations where the potential for pre-existing flaws has been indicated.

The foundation reuse decision process flow charts in *Section 5.3* identify two stages of investigation.

- *Preliminary investigations,* typically carried out before demolition of the previous structure to identify ground and groundwater conditions at the site and, where accessible, to examine the existing foundations proposed for reuse. At this stage usually only a small proportion can be checked due to limited access.
- *Post demolition investigation of the existing foundations.* Foundations will be more accessible at this stage. However, programme pressures can be significant.

The aim of the investigation of the existing piles may include:

- identification of any pile properties not apparent from the desk study review of existing foundation information,
- validation of pile properties identified in the foundation reuse desk study,
- testing for any pile defects using non-destructive methods,
- assessment of any material deterioration,
- load-bearing capacity of piles.

Checks for the more important parameters are shown in Table 3.1.

Chapter 6 gives further details of investigation of foundations.

3.3.5 Assessment of sufficiency of investigation

As identified in the foundation reuse decision process described in *Chapter 5*, the findings of the investigation of the old foundations need to be reviewed to assess whether they support the foundation reuse design assumptions. This is an important decision point in the reuse process as the strategy to reuse foundations will not be confirmed until this review is completed.

The extent of piles that need to be checked should be considered carefully. It will depend on the perceived reliability of the old records. Where the old records are uncertain but reliance is being placed on each and every one, each pile head may need to be individually exposed after demolition of the previous building to verify that they are present. However, the demolition of pile caps may in itself cause damage in the piles, which would reduce their reliability for a new design life. This and similar apparent conflicts need a higher level of judgement to balance the risks inherent in foundation reuse.

It is common in ground investigation to reveal some form of unexpected conditions such as stratigraphy or foundation geometry or conditions. Unexpected conditions may require more investigation than originally anticipated and to avoid delays associated with procuring a follow-on investigation, it is prudent to allow for contingencies.

Table 3.1 Investigation checks for key pile parameters

Relative importance	Factor	Estimation test	Parameters proven
1	Location	Positional survey Visual examination of head NDT	Pile is present Pile location Approximate pile diameter Typical materials
2	Length	Transient dynamic response (soft hammer) test Geophysics Coring	May prove length May find defects
3	Material quality/integrity	Sampling and laboratory material testing In-situ material testing	Confirm material properties Show material degradation
4	Geotechnical capacity	Load test	Geotechnical capacity (to failure?) Pile integrity

3.3.6 Foundation reuse design philosophy

For foundation reuse, the available foundations are fixed and cannot be altered. Therefore, the design approach will consist of assessing the load-carrying capacity of the existing foundations, taking into account any uncertainties in the foundation by reducing the design capacity of the foundations. To assist in the consideration of reusing single piles, a parameter relating the working load capacity required for the re-engineered pile to the previous pile working capacity is helpful (Chapman et al 2002; Chow et al 2002). This 'reuse load factor', *R*, allows the load demand at each foundation point to be assessed:

$$R = \frac{\textit{New foundation load demand}}{\textit{Old foundation working capacity}}$$

The foundation capacity of the existing pile at working load could be one of the following values.

- *Original calculations* (R_o). Technology and knowledge develops over time and the theories previously used could now be considered inappropriate for the ground conditions. Original design calculation should also be checked for any errors.
- *Current theory and approach* (R_c). This value can provide information on the expected margin against failure. Sufficient information on the geometry of the pile needs to be available and requires verification or the calculated capacity could be misleading.
- *Load take-down* (R_t). A load take-down for the previous building should be undertaken based on the building dimensions and use, and is not reliant on known foundation information. The load take-down should be based on past proven loads (from an assessment of real building usage over its life), not on previously stated design values.
- *Pile test* (R_p). Pile load tests can be carried out. The information available from each type of test compared with cost and programme must be carefully considered.

For a typical example, these reuse factors are illustrated in Figure 3.2.

The original calculations may give a lower capacity than those based on current theory and approach due to improved understanding of ground performance, validated by a database of static pile load test results. However, piles constructed before 1980 were constructed more slowly than normally happens today and therefore may have had lower capacities than similar piles constructed using current methods. Tripod bored piles in particular were built slowly and sometimes with greater seepages into the bore than might happen with modern construction methods.

If the foundation capacity determined via a load take-down of the previous structure is higher than that calculated on the basis of the pile geometry and ground conditions using current practice, there could be cause for concern as the original building was loaded higher than new calculations would allow. This is normally related to the foundation operating at a lower factor of safety than would be desirable, but still performing adequately. This may not be unusual for old buildings, whose foundations were designed empirically, and where the dead load was often added over a period of time and the building was constructed with more flexible materials. In this situation, foundation reuse would be possible if it is believed that the old structure worked well and

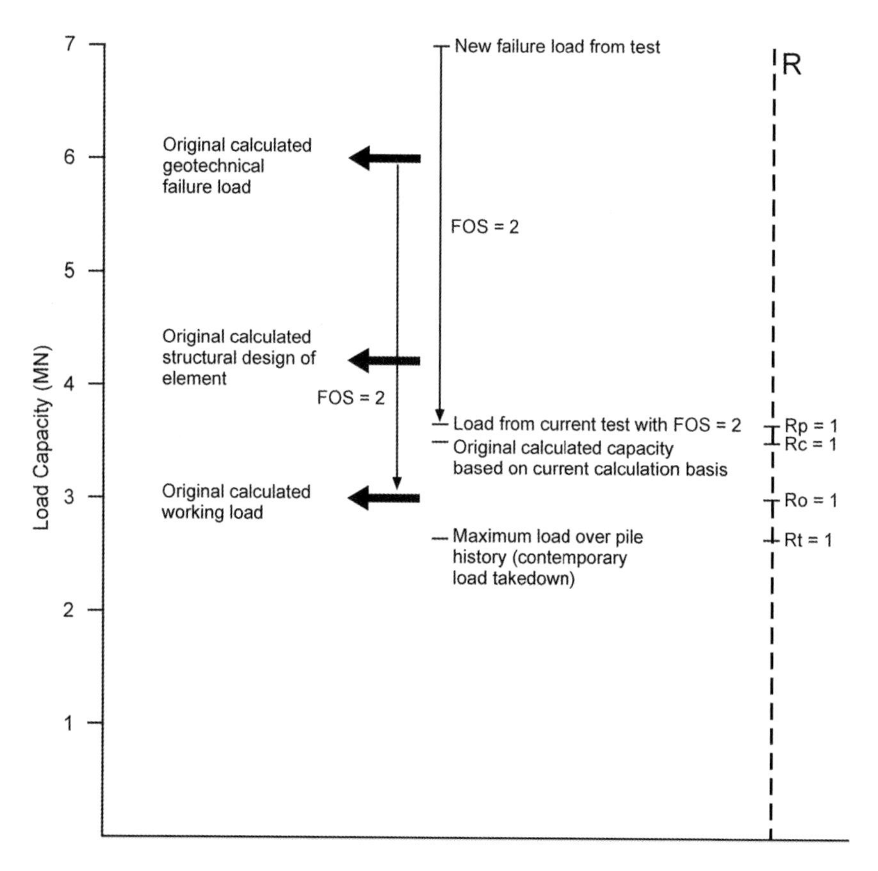

Figure 3.2 Example illustrating relation of different reuse load factors (*R*) for single piles

excessive settlement did not occur. An assessment of possible advantages and disadvantages of the different reuse load factors, *R*, is given in Table 3.2.

In many cases to date, the reuse factor based on load take-down (R_t) has been found to be the most useful calculation approach.

Assessment of foundation capacity is further discussed in *Chapter 6*, including issues relating to the potential for reserve capacity.

3.3.7 Evaluation of structural strategy for new building

The combination of foundation reuse desk study and foundation investigation may not confirm the foundation reuse design assumptions. This may result from:

- incompatibility between desk study records and findings of the investigation,
- defects identified in the piles,
- material deterioration.

Where this occurs several options for the structure are possible.

- Design 'around' clearly identified defect or suspect piles. This will require confidence in the ability to identify suspect piles from the rest of the pile population.
- Proceed with foundation reuse on the basis of a downgraded pile capacity. This may require some supplementary new piles in areas where the existing foundations no longer provide the required capacity.
- Design using a foundation system that incorporates redundancy (see Chapman et al 2002 on reliability of piles and groups).
- Incorporate the potential for upgrading foundation capacity (see *Chapter 7*).
- Design a sufficiently stiff structure to transfer loading away from potentially overstressed pile towards piles with reserve capacity.

The design solution consisting of one or more of the above options will be heavily dependent on site-specific details such as the distribution of loading in the building in comparison with the available foundation capacity.

3.3.8 Control of construction of foundation system for new building

During the installation of any new foundations to supplement the reused ones, and the construction of the new pile caps and ground slab, vigilance is required to make sure that critical design assumptions have been met. This is the stage when most of the reused foundations will be visible, unlike earlier stages when only small samples can be checked. Generally, changes in the reuse philosophy at this stage will be very disruptive, but if data become available that indicate that the reused foundations may not be sufficiently reliable, urgent measures will need to be taken.

Depending on the intensity of reuse, the reliance being placed on individual old piles and whether the old piles will have been exposed (sometimes the reuse takes place over the old pile caps, so the old piles cannot be seen), it is recommended that a number of critical piles are checked to ensure that they comply with the design assumptions. Thus, this step offers a last chance to confirm the philosophy, and to provide extra capacity if needed. At later stages of construction, the installation of replacement foundations will be difficult, costly and time-consuming.

3.3.9 Monitoring of the new building

Monitoring of the completed building on the reused foundations can play a part in the assurance strategy for the project. Monitoring needs to have started before the main loads are applied so the monitoring points should be installed and background readings taken before construction of the new building.

Although foundation movements will generally be small, in many circumstances they may provide useful data on the likelihood of cracking or other damage in different parts of the building. Having the monitoring in place from an early date should allow the cause of any damage to be quickly

Table 3.2 Features of different calculation approaches for reuse load factor, *R*

Calculation approach	Reasons why approach might be valid	Reasons why approach might not be valid
Original calculations (R_o)	Every effort would have been made to achieve accurate load estimate at that time.	Available calculations might not have been final and may have been superseded. Current theory may show it had an inappropriate design basis. Design loads may have been too onerous (eg very high live loads and might not have been tested).
Current calculations (R_c)	Based on current understanding of soil and foundation behaviour.	Crucially dependent on construction technique and on constructed dimensions and condition of foundations, which may not be known reliably.
Load takedown (R_t)	Based on loads proven over many years of successful operation.	Must be based on proven loads, especially live loads. Unless something is known about the history of the building these can be difficult to ascertain.
Pile testing (R_p)	Based on proven load test.	Pile tested might not be representative of piled system, either in dimensions or other piles' previous loading history.

explained, saving uncertainty that could inhibit the use of the building until repairs have been carried out.

It should be borne in mind that where foundation behaviour is expected to be both stiff and brittle, early monitoring may not provide much additional reassurance.

If monitoring is chosen because of uncertainty about future foundation performance that has not been resolved properly at an earlier stage, the overall strategy is inherently more risky than normally accepted and the following principles should apply.

- The consequences of the foundations performing less well should not lead to a greater risk to people than exists in normal structural design. A structural failure may occur relatively quickly and certainly between normal monitoring intervals.
- The client must explicitly accept the risks being taken on their behalf.

3.4 Acceptable risk

3.4.1 Optimal and acceptable risk

The most common form of decision analysis is where the decision criteria are used to minimise the expected cost (or maximise the expected utility). For a cost-based analysis, the decision alternative that has the lowest total cost is the optimal choice, ie is the alternative with the optimal risk.

However, there are regulations (eg Eurocodes) that specify which risk level is acceptable to society and which must be followed. This is concerned with ensuring public safety and making sure that the construction of public projects is sufficiently reliable.

The principle of optimal and acceptable risk is illustrated in Figure 3.3. On the x-axis there are different decision alternatives with different corresponding costs. In the example shown, the 'optimal' solution in terms of cost would not be sufficiently robust, so a less risky but more expensive solution is demanded by compulsory structural codes. Sometimes, the probability of the uncertain cost (risk) is too high to accept. This occurs when the decision-maker is risk-averse or there is a restriction on the magnitude of the risk that society is willing to accept. In these cases, a different alternative should be chosen that reflects the preferences of the decision-maker or the acceptable risk of the society.

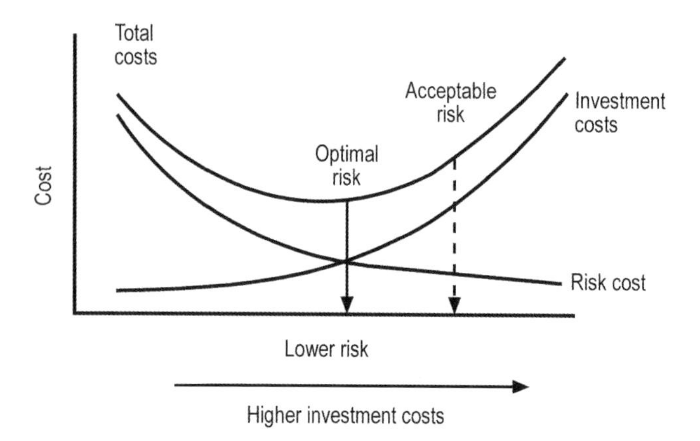

Figure 3.3 The concept of optimal and acceptable risk

3.4.2 Levels of acceptable risk

Traditionally, the concept of factor of safety (FoS) between the resistance and the action effect has been used to describe the safety level both in geotechnical and in structural engineering practice. However, in many new codes, the ambition is to prescribe an acceptable risk, or rather the acceptable probability of failure, P_f (Alén 1998; Ellingwood 2000). For example, in Sweden, three so-called safety classes are used. The safety classes have different acceptable probabilities of failure (Table 3.3).

In BS EN 1990 the acceptable probability of failure, P_f, and the corresponding target value of reliability index, β, are given for two reference periods of time: 1 year and 50 years. These are given in Table 3.4 for reliability class RC2 structures which correspond to residential and office buildings where consequences of failure are medium risk.

Further details of the definition of reliability index, its relationship with the acceptable probability and its application are documented in BS EN 1990 and Alén (1998).

The use of acceptable probability in the assessment of different foundation options is given in the idealised case study presented in Appendix B.

3.4.3 The use of caution to achieve acceptable risk

A full assessment of the probability of failure and reliability index is often not feasible for many projects due to a lack of detailed information on which to base an analysis and a lack of experienced practitioners of detailed probabilistic analyses within the building industry. In practice, risk levels are reduced through the process of risk management and through a suitably cautious approach in the design process. For foundation reuse this is typically introduced through the design load reuse factor, *R*, assigned to the foundations as defined in *Section 3.3.6*.

The value of the reuse load factor assigned to the foundations will depend on the amount of caution that is prudent to apply, for the particular use and characteristics of

Table 3.3 Acceptable probability of ULS failure depending on safety class, Swedish norms

Safety class	Acceptable probability of failure	Description of consequences
1	10^{-6}	High risk for people to be injured if a failure were to occur
2	10^{-5}	
3	10^{-4}	Very low: no risk of personal injury

Table 3.4 Target reliability index (β) and corresponding acceptable probability of failure (P_f) for reliability class RC2 structural elements from BS EN 1990

Limit state	1 year		50 years	
	β	P_f	β	P_f
Ultimate	4.7	$\sim 10^{-6}$	3.8	$\sim 10^{-4}$
Serviceability	2.9	$\sim 2\times10^{-3}$	1.5	$\sim 7\times10^{-2}$

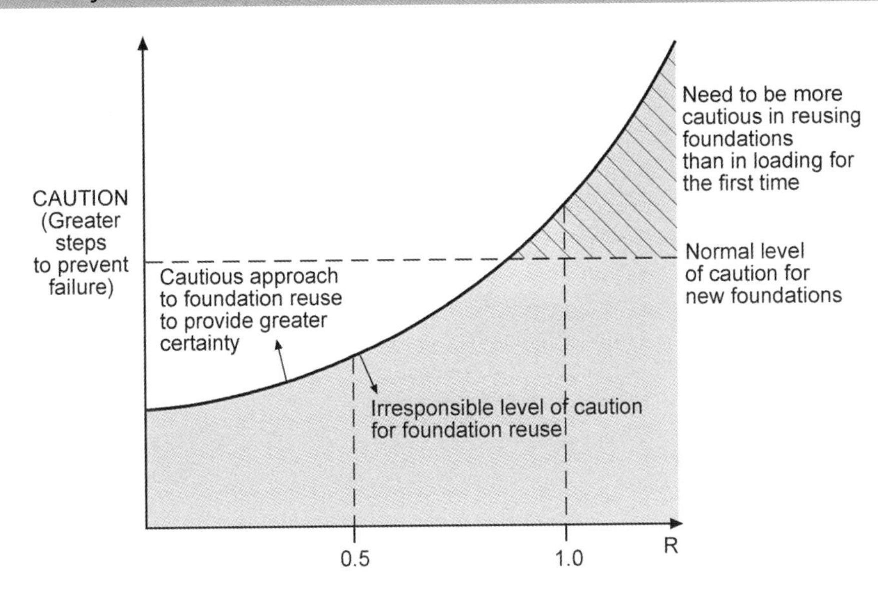

Figure 3.4 Reuse caution

the structure. For a given reuse factor, the appropriate 'caution' function will depend on:

- the robustness of the structure and its ability to share the load between adjacent foundations,
- the amount of reliable information on the foundations which is available,
- the amount of investigation that should be carried out.

A robust superstructure able to distribute loading away from a potential underperformance pile and constructed on piles with good as-built records will provide much greater caution for reloading than a less robust superstructure constructed on piles with poorer quality or incomplete records. In the latter case, more investigation of the foundations may be required to provide appropriate levels of caution for reloading. If the amount of information required becomes excessive, it will probably be more effective to supplement the old foundations with new ones, to reduce the required *R* factor.

Greater levels of caution are needed for proportionally higher applied loads on reused foundations. If the applied loads are significantly lower than those that acted previously, less caution is needed as illustrated in Figure 3.4.

3.5 Special design case: seismic loads

There have been many notable foundation failures following large earthquakes, examples of which are listed below.

- The liquefaction-induced bearing capacity failure of the Kawagishio-cho apartment buildings following the 1964 Niigata earthquake.
- The complete shearing of piles supporting a warehouse on Port Island due to lateral ground movements during the 1995 Kobe earthquake.
- The collapse of the Government Hill School in Anchorage, Alaska, which straddled the scarp of the Government Hill landslide in the 1964 Alaska earthquake.

It is clear from these cases and similar ones that these foundations would not be suitable for reuse. However, it is also clear that appropriately designed and constructed foundations can and have performed adequately in many earthquakes.

Piles are used in seismically active areas where soft or liquefaction susceptible soils may cause considerable shaking or ground displacements that can lead to severe damage of the structures they support. Foundation reuse offers considerable opportunity for the redevelopment of sites where the superstructure has been damaged by an earthquake. However, caution must be exercised when reusing them due the greater potential for defects in the foundations which could lead to long-term settlement or foundation instability.

Causes of failure of piles supporting buildings

Following earthquakes, piled foundations may suffer failure as a result of:

- *excessive bending moments*. This phenomenon may be caused for example at interfaces between different soil types such as between liquefied and non-liquefied soil, where there is a significant difference in the lateral deformation of the soil during the earthquake.
- *buckling*. This phenomenon is particularly important for structures above water or where a deep liquefiable layer may be present.
- *mass ground movement*. This phenomenon could be a result of the liquefaction causing lateral spreading, or due to a landslide.
- *the pile-cap connection,* especially where raking piles have been used.

Key issues for foundation reuse

- Existing piled foundations will usually have been designed to old seismic design codes, or possibly to no code, depending on their age, and their location. There will therefore be a requirement for detailed analysis of their expected performance under earthquake loading. If there is poor information, the uncertainties and the consequences may be too great to consider reuse. The pile design should consider the requirements of Eurocode 8 (EN 1998: Parts 3 and 5) and FEMA (2000).

- Piles that have experienced seismic loading due to earthquakes may have incurred damage that could affect their performance. This could be at depth (eg at the interface between soft and stiff layers) and therefore be impossible to inspect so will require intrusive or non-destructive testing as outlined in *Chapter 6* to confirm the condition of the foundation.
- Acceptable foundation deformations in the case of earthquake loading in terms of tilt, settlement, or differential distortion are not well defined in codes or other literature and will require judgement and client agreement as to their acceptability. Guidance is provided in EN 1998: Part 3, FEMA (2000) and PIANC (2001).

Upgrading foundations in seismically active areas to new standards
If a design check indicates that the piles are not considered acceptable, mitigation measures are needed. Methods of mitigation include:
- soil improvement of the surrounding soil,
- supplementing the existing foundations with new piles to ensure the overall capacity is acceptable.

Whatever the solution, it is important for the design to consider all the possible earthquake hazards described in EN 1998: Part 5, which include:
- seismically active faults,
- landslides,
- liquefiable soils,
- settlements of soils under cyclic loads.

3.6 Key points

The reuse of foundations increases the potential risk of failure of a redevelopment in terms of:
- building performance,
- structural failure,
- construction costs,
- construction programme.

The sources of additional risks associated with foundation reuse are predominantly concerned with ways of over-estimating the load-carrying capacity or resistance of the foundations to be reused including:
- absence of foundations due to major design changes,
- presence of major pre-existing flaw in the foundations,
- poor load-settlement performance of old foundations,
- application of eccentric loading to old foundations,
- loss of durability in the foundations,
- occurrence of seismic events

However, if these risks are clearly identified and managed by means of a thorough risk management process introduced in *Chapter 5*, including key stages of risk reduction:

Identification of potential risk through:
- appraisal of old building operation,
- foundation reuse desk study of available information and knowledge,
- investigation of old foundations.

Assessment of identified risks through:
- assessment of sufficiency of investigation,
- foundation reuse design approach.

Control through:
- evaluation of structural strategy for the new building,
- control during construction of foundation system for new building.

Monitoring:
- monitoring of new building to confirm design performance

The use of this approach will increase the likelihood of achieving the potential benefits of foundation reuse including:
- reduced construction cost,
- improved construction programme,
- reduction in environmental impact.

3.7 References

Alén C. On probability in geotechnics. Random calculation models exemplified on slope stability analysis and ground-superstructure interaction. Volume 1. Dissertation. Department of Geotechnical Engineering, School of Civil Engineering, Chalmers University of Technology, Göteborg. 1998

British Standards Institution. BS EN 1990: 2002 *Eurocode. Basis of structural design*

British Standards Institution. BS EN 1997-1: 2004 *Eurocode 7. Geotechnical design. General rules*

British Standards Institution. BS EN 1998: *Eurocode 8. Design of structures for earthquake resistance*
Part 3: 2005 Assessment and retrofitting of buildings
Part 5: 2004 Foundations, retaining structures and geotechnical aspects

Chapman TJP. Ground Engineering Talking Point. *Ground Engineering* 2004: **June**

Chapman TJP, Chow FC & Skinner H. Building on old foundations: Sustainable construction for urban regeneration. CE World Conference, ASCE on-line conference. 2002

Chow FC, Chapman TJP & St John HD. Reuse of existing foundations: planning for the future. *Proceedings of 2nd International Conference on Soil Structure Interaction in Urban Civil Engineering,* Zurich, 2002

Ellingwood BR. Probability-based structural design: prospects for acceptable risk bases. In: Melchers RE & Stewart MG (eds) *Applications of Statistics and Probability,* Rotterdam, Balkema, 2000

Federal Emergency Management Agency (FEMA). Prestandard and commentary for the seismic rehabilitation of buildings. FEMA 365, November 2000

PIANC. *Seismic design guidelines for port structures.* Rotterdam, Balkema, 2001

Simpson B & Driscoll R. *Eurocode 7: a commentary.* BR 344. Watford, IHS BRE Press, 1998

St John HD & Chow FC. Follow these footprints. *Ground engineering* 2001: **December:** 24–25

4 Legal and financial context

4.1 Introduction

Foundations are just one of many elements of a completed building. While they are a relatively small part of the overall cost, they constitute a relatively large part of the overall project risk. They lie on the project critical path: few other activities can start until they have been completed. The consequence of a foundation defect may affect large parts of the structure they were intended to support. Repairs to the foundations or the installation of supplementary foundations are difficult and disruptive. Foundations therefore need to be seen in the context of the overall project, and the special features of reused foundations also need to be considered in that context.

As the foundations for any project are so important for the rest of the project, they need to be procured based on more than just their direct cost. The crucial aspects that need to be considered are:

- overall cost for the foundation system (eg not just direct costs, but also other substructure items such as pile caps),
- programme for installation,
- risk that either cost or programme will be exceeded during installation,
- risk of defect or failure that needs investigation and rectification, including disruption to other construction activities and programme over-run, possibly delaying occupation,
- risk of failure that only becomes obvious after occupation, causing disruption to users of the completed building.

For most construction clients, the highest priority is to avoid construction problems. A second priority is to make sure that risks are 'owned' by those best able to control them, ensuring that if a problem does occur there is someone who can be held liable and ideally from whom losses can be recovered. *Section 4.2* addresses likely responsibilities should a foundation reuse project go wrong. It concludes that there are various strategies to offset financial risk for a client, but these cannot be relied on to be completely successful.

The assessment of risks and comparison of options to produce the lowest overall risks can be carried out qualitatively by the design team, who will use the risk reduction approach set out in *Chapter 3*. A quantitative risk analysis early in the decision process allows a more rigorous approach to be taken, provided the input data is correct and appropriate. An idealised case history is therefore presented in Appendix B and can be used to demonstrate some of the considerations that should be included in a comparison of foundation options.

4.2 Legal framework

4.2.1 Types of legal liability

For any civil engineering project, there will be a promoter/developer/funder/user who will seek legal protection for any failure by his designers/consultants or construction contractors. This section of the *Handbook* addresses the extent of legal protection that may apply for different situations.

In most European states, design consultants' liability is linked to the exercise of professional skill and care. Even if the design is flawed, the law states that the designer will only be liable if it can be shown that he failed to exercise professional skill and care. Design consultants' professional indemnity insurance is based on this measure of performance and a consultant who accepts a higher contractual obligation may find himself without insurance for any additional liabilities assumed.

In contrast, contractors' liability is usually absolute: if there is a defect in the building which stems from something for which the contractor is responsible, he will usually be liable even if he can show he took reasonable care to avoid the problem. This is often known as 'fitness for purpose' liability.

Due to the inaccessible nature of foundations, it may be difficult to distinguish whether a foundation failure is due to inadequate design or poor performance.

4.2.2 Procurement routes

The two types of procurement route which are most relevant to building development projects are:

- contractor design and build, and
- employer design (through consultants) with contractor building (sometimes known as 'traditional'; this term is used below).

In *contractor design and build*, the contractor takes complete responsibility for the finished product (although design is often delegated by the contractor to consultants who are employed as subcontractors). A hybrid prevalent in the UK involves the contractor accepting responsibility for the consultant team which has been initially engaged by the employer ('Novated design and build'). In *traditional procurement,* the employer pays consultants to prepare designs which potential contractors tender against. The selected contractor then builds what the consultants have designed.

Applying the different types of legal liability identified in *Section 4.2.1* to these alternative procurement routes, the risk transfer can be compared.

When advising employers on foundation reuse options, the consultant must consider these issues in conjunction with the technical assessment, having regard to the employer's appetite and capacity for risk.

Traditional procurement

Employers who engage consultants at the initial stage to investigate the existing structure and to advise on options need to understand the following points.

- A consultant who exercises professional skill and care in relying on historic foundation data (eg as-built drawings) may not be liable if the drawings prove to be incorrect, unless it should have been obvious that they could not be relied on (eg they show a different column grid to the one in the building).
- Similarly, a consultant who commissions a site investigation which is shown later to have failed to detect adverse evidence relating to existing piles, will not be liable if he has exercised professional skill and care in planning and interpreting the investigation.
- A consultant who designs a foundation reuse scheme, making reasonable professional assumptions about the existing foundations, will not be liable if those assumptions prove to be incorrect.

In traditional procurement, some design risk is retained by the employer. For example, if the contractor encounters an unforeseeable problem with ground conditions on site which requires a change in design, the costs of redesign and of the attendant variation to the works, together with the delay-related costs, would be borne by the employer. In each case, the costs of redesign and rework, together with knock-on costs associated with delay, fall to the employer.

Design and build procurement

The decision as to whether to reuse foundations may be taken by the design and build contractor rather than the employer, whose ability to secure a particular design methodology is restricted by the fact that the employer's design input is limited to a set of outline performance requirements provided to the contractor at an early stage.

If the contractor employs consultants, then in the event of a problem he will face similar defences from the consultants to those outlined above; however, he is unlikely to be able to raise equivalent defences with the employer. A well-advised contractor will recognise this risk and will price accordingly, resulting in the employer paying a premium. The employer will seek to pass delay risk to the contractor using liquidated damage provisions.

The risks for the employer are less obvious than for traditional procurement, but are there nonetheless, ie losses associated with delay may not be covered by liquidated damages (eg if a prospective tenant terminates his agreement for the lease), and the small but ever-present threat of contractor insolvency raises the possibility that the risks may eventually revert to the employer.

It should be borne in mind, therefore, that risk transfer is rarely absolute: insolvency of the contractor, or flaws in contract documents, undermine the risk transfer process, and in practice, even with design and build procurement, the employer always retains a small but significant level of risk. Where a problem occurs, even a design and build contractor may believe that his brief or ground investigation reports on which he relied were inadequate, so the blame should lie elsewhere.

4.2.3 The client decision-making process

The client decision-making process will be influenced by the type of development, ie commercial or one-off development for an owner-occupier.

Commercial developments

Decisions about foundation reuse will be driven as much by commercial factors as technical ones. Consultants need to be aware of the stakeholders in any commercial development:

- developers,
- funders,
- future tenants/users.

Each has different priorities, which will be measured predominantly in terms of cost, time and building quality (although increasingly sustainability is also a factor).

Commercial developers often target rapid construction: once the building is sold or let, their profit can be realised. Any 'extras' over and above the agreed budget will generally come out of their profit: they do not want surprises. An equivalent priority is that they want to avoid any risk which might put off potential tenants/purchasers. When foundation reuse is being considered, it is usually the developer who makes the key decisions about proceeding but will do so in the consideration of what the funder and eventual asset owner would wish. If the building is bespoke towards an eventual end-user, they may also be consulted and will possibly retain their own professional advisors to protect their interests.

Funders are invariably risk averse: they usually prefer the most risk-free technical options. They are not necessarily interested in the quickest or cheapest construction but they do require certainty. They are looking for reliable long-term returns and increasing asset value in return for their investment.

Tenants/users target building quality judged over the term of their occupation. They are often advised of the risks of latent defects occurring during their occupation and they often have a perception that this is a key risk for which insurance is required (see *Section 4.2.1*). Blue-chip companies may have environmental/sustainability agendas that need to be recognised.

One-off developments for owner-occupiers
Owner-occupier developers build for their own use. They use the building as part of a broader business purpose and the building must allow them to undertake their business in an efficient and uninterrupted way. Their priorities are often consistent with those of tenants/purchasers and, having no external funders, they may be less risk-averse than commercial developers. Future options and maintaining the value of the site for redevelopments may be important.

4.2.4 Likely client attitude to foundation reuse: a summary

Against this background it can be concluded that a commercial developer may be reluctant to consider the reuse option, unless it will clearly bring significant savings if successful. A traditional procurement option leaves the developer retaining significant risk; design and build procurement may go some way to alleviating the developer's concerns by passing a majority of risk onto the contractor, but a well-advised developer will be aware that such transfer can be illusory. Even if a developer is keen to explore the option of foundation reuse, perhaps because of potential programme savings, funders may not be so keen unless again there is a tangible benefit for them in terms of reduced investment without affecting returns. The maintenance of better records (as later advocated in *Chapter 7*) should do much to alleviate these risk concerns and make foundation reuse more viable in the future. Better records of what lies buried in the ground should give more confidence to designers of reused foundations, and hence to their clients. This is the step-change in behaviour that will make foundation reuse more feasible, even in situations where its direct benefits are marginal.

The concerns of tenants/purchasers may be easier to address, for example by using latent defects insurance, also known as decennial insurance. And if the project is not speculative, but the developer is undertaking it specifically with a blue-chip tenant contracted to take a lease, the tenant may find attractive the sustainability arguments associated with reuse.

A one-off development for an owner-occupier may be more likely to employ foundation reuse. With fewer stakeholders, the owner-occupier may be more inclined to assume the risks associated with foundation reuse, may be more willing to commit greater funds to up-front investigation, and may be able to manage the risks associated with project time or cost over-run. The blue-chip owner-occupier may also find the sustainability case attractive.

4.3 Financial context

4.3.1 General financial issues

Every project where reuse is an option is unique, with its own special set of constraints, demands and participants' views. There can therefore be no simple financial rules, apart from the following broad observations.

- Clients adverse to risk and delay may have a bias against reuse.
- The level of risks being taken should be similar to the level associated with the installation of new foundations.
- Foundation reuse is made much more feasible if good information is available on the old foundations (see *Chapter 7*).
- Foundation reuse tends to be broadly welcomed if it makes tangible savings in cost or programme, and the extra investigation costs and perception of risk does not detract from these savings.
- Whole life costing may bias the solution either in favour or against foundation reuse, and should always be considered.
- Protection of archaeological resources can be a strong driver towards foundation reuse and can reduce the viability of a development incorporating new foundations.

4.3.2 Specific financial issues

Chapman & Marcetteau (2004) identified the cost distribution for a typical UK building project as shown in Figure 4.1.

For the typical building type identified, the site investigation costs are usually just 2% of the substructure costs or 0.1% of the overall building cost despite ground-related problems typically accounting for about one-third to one-half of construction programme over-runs.

In this case, foundation reuse offers the potential to reduce the project costs by eliminating a significant proportion of the £600K cost of foundation construction. However, this reduction in cost must be balanced by increased investigation costs associated with foundation reuse but also the increased risk of failures associated with any unknown defects in reused piles.

Appendix B shows a case study for decision making based on financial risks that includes the impact of programme disruption.

4.4 Key points

Foundation reuse has the potential for savings in both construction cost and programme, balanced by a potential increase in risk of foundation failure associated with reusing existing foundations. Attitudes to foundation reuse will be affected by the project procurement route as well as by the different parties involved depending on their different drivers and appetites for risk.

- Commercial developers operating with a traditional procurement route are likely to be risk-averse, preferring increased cost and programme certainty, often at a premium. Design and build procurement may alleviate

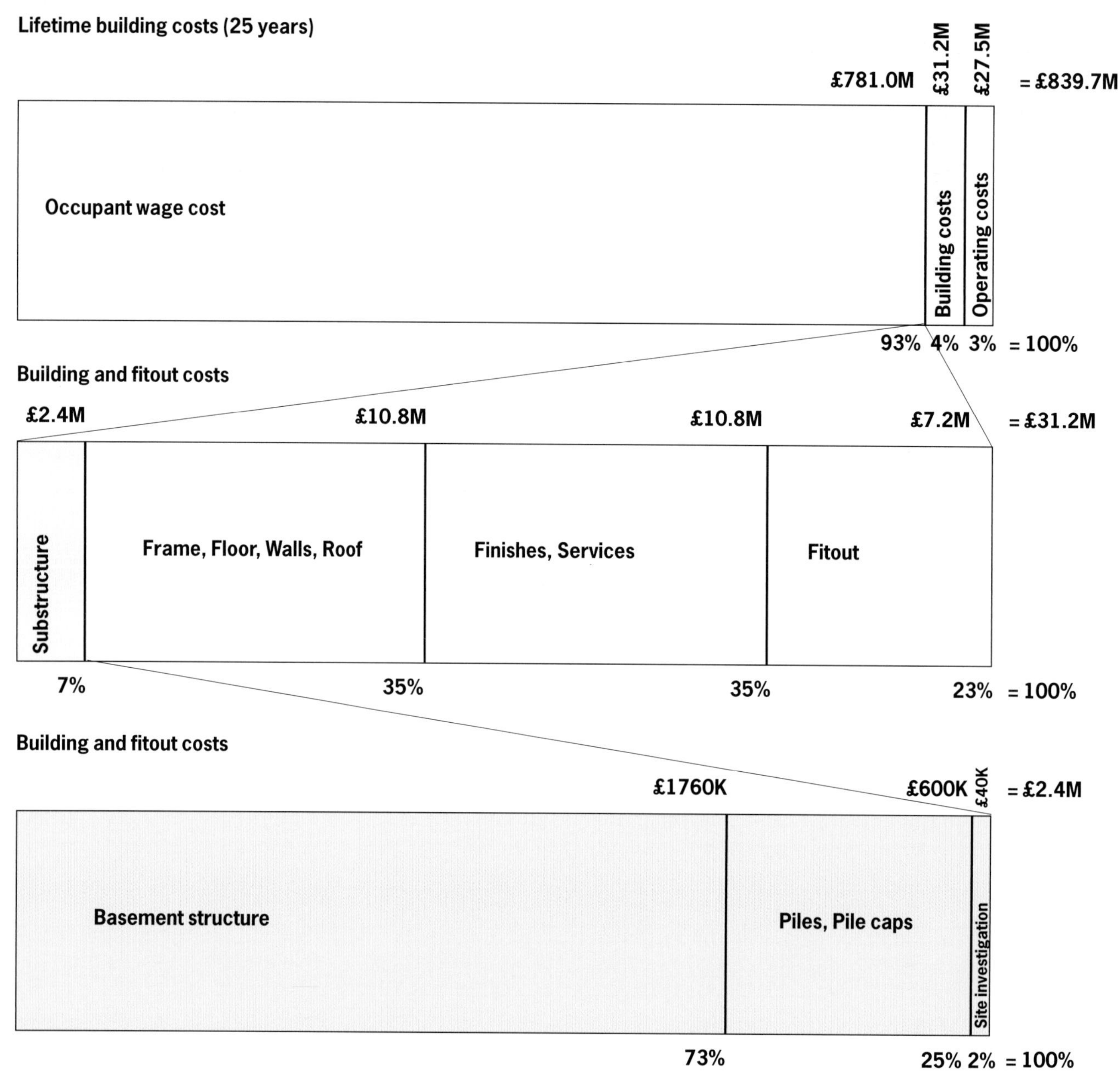

Figure 4.1 Lifetime building costs (Chapman & Marcetteau 2004)

some of the developer's concerns by appearing to pass risks onto the contractor.

- Tenants/purchasers' concerns for the risks associated with foundation reuse may be addressed through latent defect insurance, also known as decennial insurance.
- A one-off development for an owner-occupier may be more likely to employ foundation reuse and manage the risks during the design including investment in early investigation to reap the potential benefits, in particular in construction time and sustainability benefits.

Through use of an idealised case study comparing different foundation solutions by means of a Bayesian network analysis (see Appendix B), several features of foundation reuse have been demonstrated.

- For foundation reuse schemes, the probability of failure due to defects in the existing foundations is reduced for foundations where good quality records are available and the perceived status of the existing foundations is good (based on judgement of the original installation, records and previous performance).
- Structural redundancy in the foundation system assists in the reduction of failure risk for the foundations due to potential defects in the existing foundations: in this case, redundancy was provided through multiple piles under a single cap rather than a single, albeit larger, pile individually supporting each column location. This approach is slightly at odds with a frequently perceived programme benefit of eliminating pile caps by choosing single piles per column.
- When assessing the different foundation options relevant to a redevelopment project both cost and likelihood of failure should be considered; a consideration of cost alone may result in a foundation option which does not provide a suitably low probability against failure, in accordance with BS EN 1990. In practice, consideration of probability of failure is unlikely to be undertaken explicitly but will be introduced into the foundation design through suitable

levels of investigation and caution adopted in the choice of a reuse load factor for the existing foundations.

- When comparing different foundation options, costs of the foundation construction should not be considered in isolation but should be put into a larger context. For the idealised case study in Appendix B, the different assumed levels of benefits associated with programme duration, construction safety and sustainability benefits had a significant effect on the total expected cost of the foundations.

Foundation reuse has the potential to offer financial savings to a project, in particular when good quality records are available and benefits such as construction programme, construction safety and sustainability are considered. The idealised case study demonstrates some of the benefits and limitations of foundation reuse. However, these conclusions are based on the assumptions made and therefore are not necessarily valid for all construction redevelopment projects. A site-specific assessment should be made for each project to identify the particular financial risks and opportunities that foundation reuse may offer.

4.5 References

British Standards Institution. BS EN 1990: 2002 *Eurocode. Basis of structural design*

Chapman T & Marcetteau A. Achieving economy and reliability in piled foundation design for a building project. *The Structural Engineer* 2004: **82**(11)

5 Decision model

5.1 Introduction

The foundations for a new building should be considered at the earliest stage of a project. The compatibility of the building load points with the pre-existing foundations should be assessed and in an ideal case the structure should be designed using the existing foundation layout. This radical departure from the traditional approach will maximise the reuse of foundations and hence sustainability.

There are several foundation options available for new developments on sites with existing foundations. These are shown in Figure 5.1 for deep foundations.

The reuse of foundations is already accepted practice for nearly all refurbishment projects. However, the scope of foundation reuse considered in this *Handbook* extends to redevelopment, and is not restricted to refurbishment projects where the configuration of the new loading is generally similar to the original design loading.

Option 1: Complete reuse

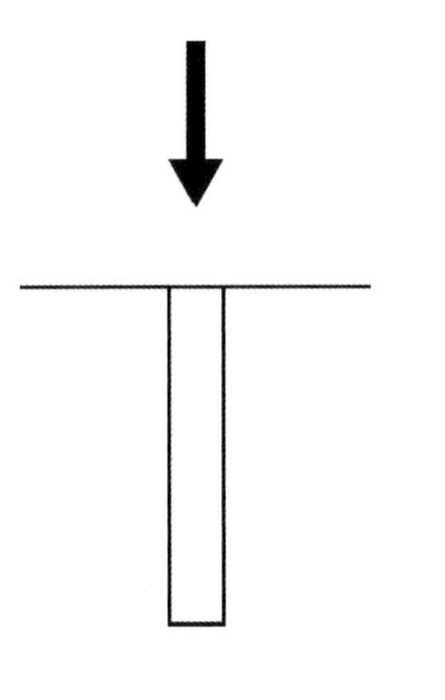

No new foundations are required but some supplemental techniques as identified in *Chapter 6* may be incorporated to improve the load-bearing capacity of the foundations

Option 2: Partial reuse

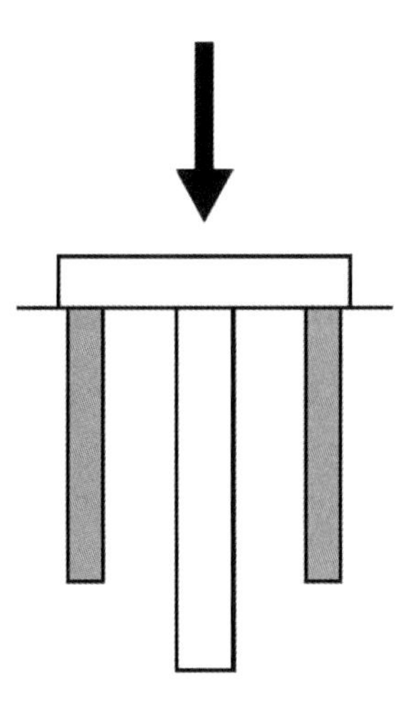

New foundations are constructed to supplement the existing foundations. (*Note:* in the case of existing deep foundations, partial reuse may also consist of a hybrid solution of constructing a new raft over existing piled foundations to create a piled raft)

Option 3: Complete replacement

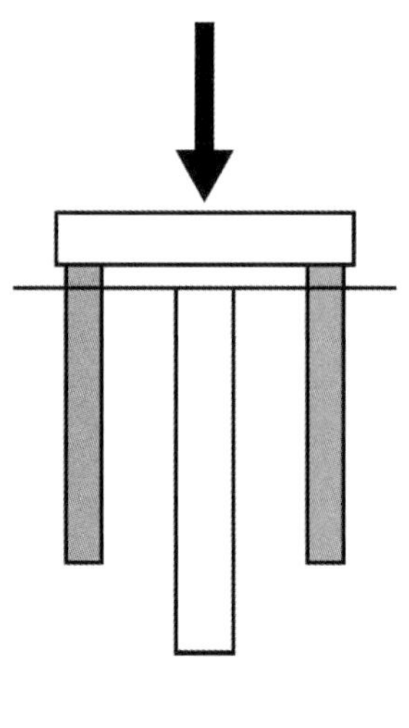

Install new foundations avoiding existing foundations

Option 4: Remove and replace

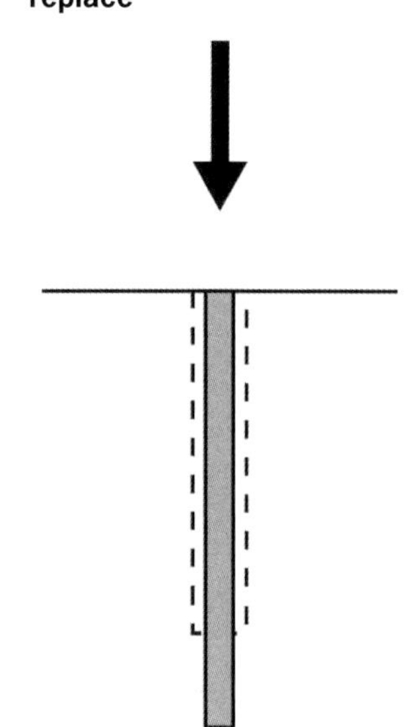

Existing foundations are removed and new foundations constructed at the same location. (*Note:* detailed discussion of this option is not within the scope of this *Handbook*. However, it should be noted that removal of existing piles can be challenging and hence expensive and time-consuming)

Figure 5.1 Options for new foundations

The ability to reuse foundations will depend on the technical complexities of each particular site. For foundation reuse to be successful, two basic technical principles must apply.

- There must be geometrical compatibility between the locations of the applied loads and the existing foundations, which must have sufficient capacity to carry the new loads and have settlement characteristics appropriate to the new structure (referred to as settlement compatibility in *Chapter 6*).
- Sufficient verification must be carried out so that the old foundations are shown to be as reliable as new ones.

Unless there are particular site constraints which prohibit new foundation installation such as the presence of archaeology, tunnels or other obstructions, it will often be necessary for the existing foundations to be supplemented by new foundations to form the new foundation system. Therefore, the key decision will seldom be a choice between foundation reuse or construction of new foundations, but will be a decision as to whether some reuse of foundations will form part of the overall foundation strategy for the new development.

Consideration of foundation reuse will also be influenced by the particular values of the client and project team as well as their attitude to uncertainty. For the decision to reuse to be acceptable:

- all members of the design and construction team should embrace a foundation strategy involving reuse,
- the client is prepared to consider such a strategy, and is able to obtain any suitable insurance products required to carry some of the risks,
- the approach is acceptable to the necessary authorities.

Key issues relating to foundation reuse are presented in *Section 5.2*. These key issues have been used to develop a foundation reuse decision process with the aim of enabling owners, funders, insurers, regulators and designers of proposed redevelopment schemes to evaluate the feasibility of reusing existing foundations. Flow charts for this process are presented in *Section 5.3*.

Like any other construction option considered, a foundation reuse strategy must offer advantages over other foundation strategies, in terms of the criteria that are important to the project stakeholders. Typical criteria or indicators include:

- construction cost and programme,
- whole life cost,
- environmental impact or sustainability.

There is a discussion of the measures needed to reduce foundation risk to acceptable levels through a risk management process that adopts these assessment criteria for foundation reuse in *Chapters 3* and *4*. An example of the decisions required during development of a typical office block is described from a financial viewpoint in *Section 4.3*.

5.2 Key issues (framework for reuse of foundations decision model)

The principal issues that need to be considered are identified in Table 5.1, together with the relevant situations that tend to be most favourable for a strategy involving foundation reuse to be successful. These issues are discussed in greater detail in the following sections.

5.2.1 Ground congestion

For sites where the ground is already congested due to previous foundations and other obstructions that are difficult to remove as shown in Figure 2.1, foundation reuse may improve the feasibility of redeveloping the site. Foundation reuse can also improve the feasibility of redeveloping sites where the preservation of archaeological remains is required as described in *Chapter 2* or where environmental conditions such as ground contamination reduce the feasibility of intrusive groundworks.

The process of desk study and field investigations is key to identifying obstructions (archaeological or otherwise), which will influence the design of a foundation system and hence the feasibility of foundation reuse. The use of a desk study as part of the risk reduction process is described in *Chapter 3* and details of the desk study process and field investigations are provided in *Chapter 6*. Figure 5.2 shows the decision process to deal with ground congestion including the preservation/avoidance of archaeology and existing foundations.

5.2.2 Foundation information

Detailed and reliable information on the existing foundations (type, location, diameter, length, reinforcement) can be crucial for efficient reuse of old foundations, in particular, where the design loading is comparable or in excess of the original building loads. This information should be gathered as part of a desk study process and confirmed by field investigations, especially where there is any doubt as to the quality or accuracy of the available information. In practice, obtaining design and construction records can be difficult as a full set of records may not have been kept or the original design and construction firms may not be in operation any more.

5.2.3 Geometric compatibility

It is often not possible for the layout of a new building to be constrained to the column grid from the previous building and also be able to fulfil economically its new function. Where the new building layout is mostly compatible with the old column grid, it may be possible to reuse the existing piles supplemented by additional new piles.

Where the compatibility is limited, it may be possible to improve the potential for either full or partial reuse by modifying the proposed building layout to suit the existing foundation layout better. However, the implications of this in terms of efficient operation of the structure or the requirements for transfer structures should be considered. The layout of a building which may be operational for a

Table 5.1 Factors governing the reuse of foundations

Key issues affecting potential for foundation reuse		Favourable conditions for reuse	Section
Technical issues	Ground congestion	Congested ground restricting the locations for the installation of new piles	5.2.1
	Foundation information	Reliable information is on the existing foundations available and relates to as-built rather than design conditions	5.2.2
	Geometric compatibility	Points of load application are close to locations of existing foundations	5.2.3
	Foundation capacity and performance	Capacity of the existing foundations is high in comparison with the applied loads and the settlement performance of the foundations is compatible with the new structure	5.2.4
Legal issues	Safety	Site conditions introduce safety risk to piling options. However, intrusive validation investigations do not increase safety risks	5.2.5
	Risk and liability	Client is made aware of the risks associated with foundation reuse, is willing to accept the risk and is able to obtain suitable insurance	5.2.6 Also *Chapter 3*
	Regulatory approval	Regulatory approval bodies are amenable to foundation reuse	5.2.7
Criteria for assessing risks and benefits of foundation reuse	Construction cost and programme	Cost and programme implications of requiring a new set of foundations are large in comparison with the implications associated with intrusive validation investigations and any transfer structure requirements	5.4.1 Also *Chapters 3* and *4*
	Whole life costs	High costs associated with the construction of new foundations and no increases in operational costs associated foundation with reuse	5.4.2
	Environmental impact or sustainability	Environmental impact of new foundations is high due to the site location close to environmentally sensitive areas or far from material suppliers	5.4.3

design life of 50 years must suit the function of the building. It is therefore unlikely that it will be economical to modify the building layout significantly to suit the existing foundation layout as the potential savings in construction costs would be more than offset by increased operational costs. Evaluation of the whole life costing is described in *Section 5.4.2*.

Figure 5.3 compares the column layout for a new building overlaid by the previous foundation layout, for a proposed redevelopment in central London. In this case, the new columns and the previous foundations only coincided at about 10% of the locations. As there were no other drivers providing benefit from a foundation reuse scheme, it was considered unsuitable and a new set of deep foundations was constructed.

If the layout of a new building is on the same grid as a previous structure and it is not feasible to remove the previous foundations, reuse may be beneficial in order to reduce the requirements for transfer structures. Figure 5.4 shows how the congestion due to existing deep foundations restricted the locations for a new set of foundations. Unfortunately, for this particular site it was not feasible to reuse the existing foundations. The need to avoid the existing pile locations resulted in additional design effort to produce many different pile and pile cap configurations across the site to support the building columns.

In rare instances, it will not be desirable to reuse the old foundations and new foundations must be placed at the same locations. This can occur on archaeological sites where higher capacity new foundations are constrained to the same locations as the old foundations installed in less archaeologically aware times. Another example is where a new embedded pile retaining wall is constrained to run through an old pile group. For these examples, the existing piles may need to be removed and replaced by a new pile at the same location.

5.2.4 Foundation capacity and performance

The reuse of existing piles is already accepted practice for nearly all refurbishment projects. For these projects, the configuration of the new loading is generally similar to the original design loading and foundations have been allowed by some authorities to receive an increase in loading: eg in the UK BCA/BSRIA (Gold & Martin 1999) state:

'Extra load capacity is generally built into the foundations of older buildings and in the 1970s Building Control would commonly allow an additional 10% of the total building loading to be added to the foundations at a later date, provided the building was sound and no settlement had occurred.'

For foundation reuse the existing foundations, in conjunction with any supplementary new foundations, must have

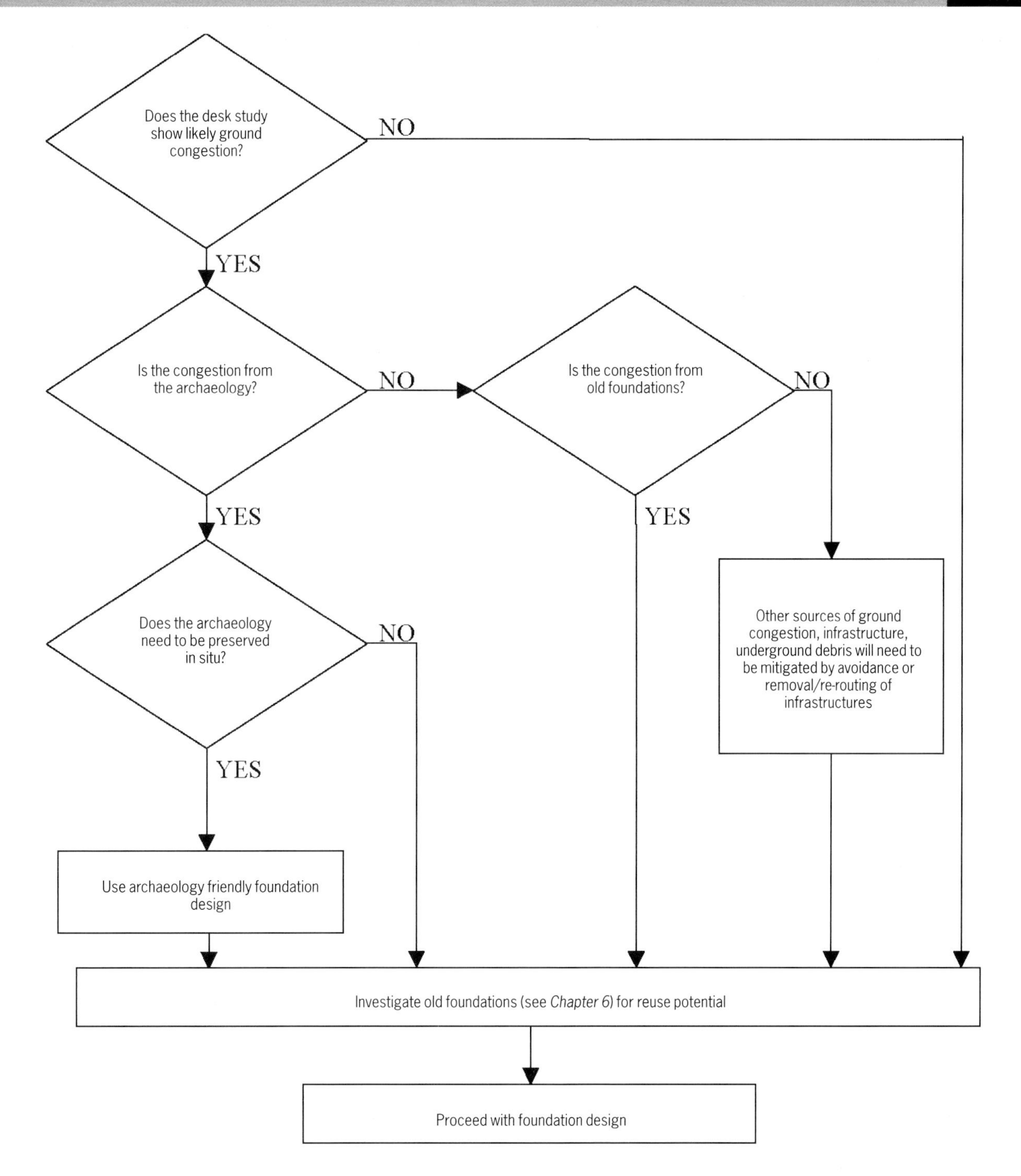

Figure 5.2 Decision process to deal with congested ground

sufficient capacity to carry the new design loads safely. To assess the capacity of the existing foundations, sufficient information on the old building and the sub-structure is required. If an unduly conservative estimate is made (to compensate for limited foundation information), the foundations will be working below optimal capacity so the new building design will be less efficient and hence less economical than otherwise possible. A conservative assessment will also increase the likelihood that supplementary foundations will be required to provide sufficient load-carrying capacity. It may be possible to identify hidden reserves in the existing foundation system or to incorporate supplemental methods to improve the foundation load-carrying capacity as described in *Chapter 6*.

It is also important to confirm that the foundations performed satisfactorily under the loads previously believed to be imposed. If the building exhibited signs of damage or there is evidence or records of poor performance or structural repairs having taken place, it may be necessary to adopt a more cautious approach to foundation reuse (eg reduce the design reuse load capacity), or even discontinue consideration of foundation reuse.

The assessment and design of existing piles for reuse is presented in further detail in *Chapters 3* and *6*.

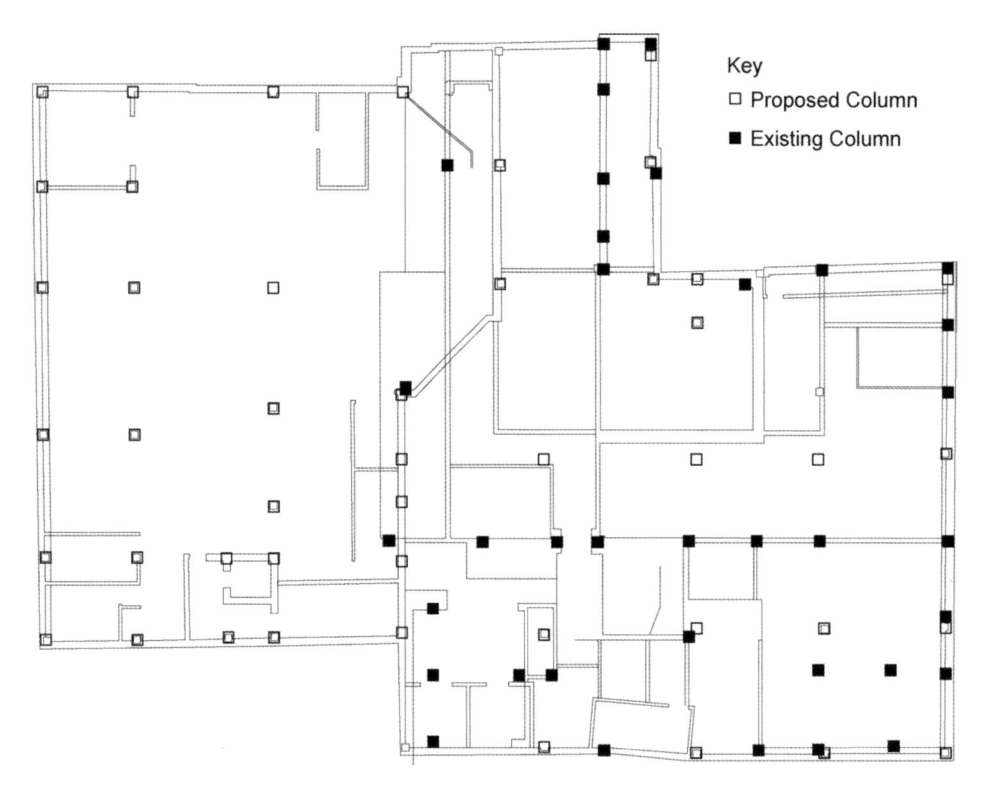

Figure 5.3 Incompatibility of proposed and existing column locations

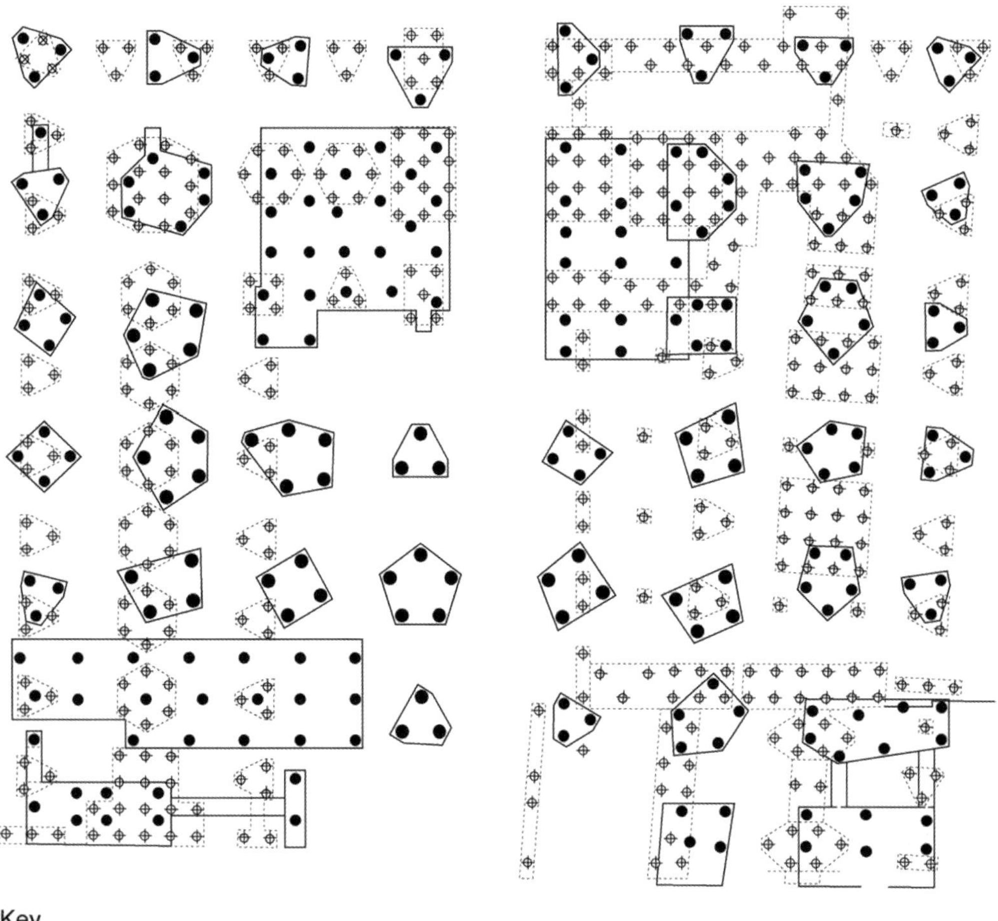

Figure 5.4 Ground congestion from a first set of piles limiting locations for a second set of piled foundations

5.2.5 Construction safety

On all construction projects, the reduction of risk to health and safety is of paramount importance. Legislation at a European and national level ensures that safety is a key consideration in the choice of techniques for construction projects. Additionally, the costs of accidents on building sites are completely disproportionate to any small savings that may have been made so maintaining cost and programme should also mean that keeping a safe site is a main driver.

In all comparisons between different foundation options, the maintenance of site safety and the protection of human health must therefore be kept for all options, and comparisons should not be based on possible savings by compromising health and safety standards. In *Chapter 4*, the cost implications of less safe strategies are examined. This is not because the principle of using a less safe strategy is endorsed; *Chapter 4* merely seeks to demonstrate that less safe strategies have significant cost implications which make them undesirable.

Where an option potentially carries a higher risk, the risk should be flagged for later follow-up. This is a requirement of an EU directive (Council Directive 92/57/EEC on the implementation of minimum safety and health requirements at temporary or mobile construction sites), enacted in the UK in 1994 as Construction (Design and Management) Regulations (HMSO 1994) and which came into practice in 1995. Particular risks that might be higher could include those shown in Figure 5.5.

The mitigation of the risk can be done through subsequent stages of design (eg use reinforced concrete shear walls within a basement as transfer structures) or by site operations (eg grub out old piles to a level below the piling mat so that piling rigs are not affected by such hard spots).

5.2.6 Risk and liability

This section gives a brief discussion of the key risks in foundation reuse. *Chapter 6* presents design methodologies that can reduce these risks at each stage of development.

The risks associated with pile reuse are comparable with the risks associated with unforeseen ground conditions, or refurbishment of an existing building:

- uncertainties regarding what cannot be seen before construction,
- with attendant risk of remedial measures being required during construction leading to cost and time over-runs, and possible compromise of the client's aspirations for the building, and
- in extreme situations, abandonment of the scheme or even building failure.

Experienced construction professionals should be able to address and manage these risks on their client's behalf. In particular, they should be able to explain and quantify the risks to dispel misconceptions and advise on risk mitigation strategies. As with ground risk, early investment in investigation and feasibility assessment will pay dividends later.

The client's engineers must, therefore, ascertain the client's priorities and appetite for such risks at the outset. The client should also be made aware of the potential liability limits of those employed as advisors or constructors to identify the resulting liabilities which will remain with the client as listed below.

- The original designers/constructors (if they can be found) are unlikely to offer any warranty in relation to the as-built state of the original piles. Any information they offer is likely to be provided on an 'as is' basis, with no warranty.
- The client's design team will need to identify the level of information and form a view as to its accuracy and

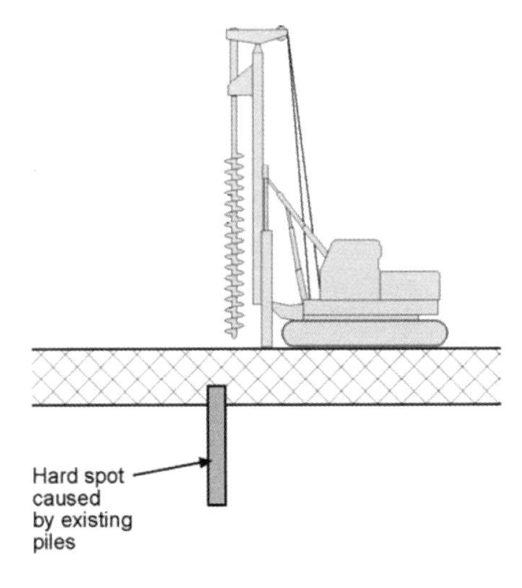

Foundation reuse and augmentation
Piling rig destabilised by tracking over existing piles which act as hard spots under piling blanket

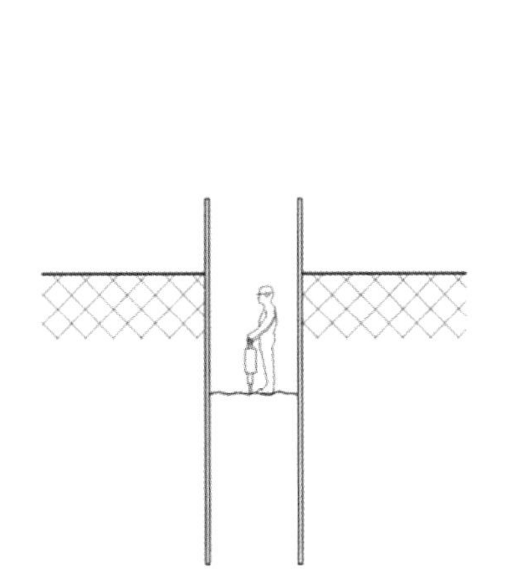

Foundation removal and replacement
Risk to worker entering open bore to attach a lifting lug onto a length of old pile that has been over-cored

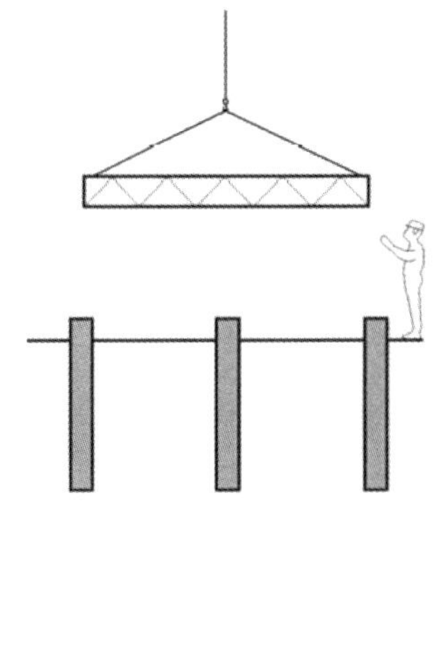

New foundation installation
Lifting risks for large steel transfer structures being craned over the site

Figure 5.5 Safety risks relating to foundations

adequacy. It is important to note that if the design team makes engineering judgements that subsequently turn out to be incorrect (eg over-estimating the capacity of existing piles), they will only be liable if the client can show those judgements were negligent.

- Unless the client has passed the risk associated with existing pile capacity to the contractor, the risk will be retained by the client.
- Seeking to pass the risk to the contractor via a 'fitness for purpose' building contract may appear to offer comfort but there is a premium to be paid and if the risk materialises it is unlikely that the client will escape loss entirely. In some situations (eg contractor insolvency) the risk will revert to the client.
- Similarly, relying on collateral warranties to give building users a right to seek redress is not necessarily an effective means of transferring the risk.

The normal contractual organisations used in the UK are illustrated in Figure 5.6. Similar organisations are present elsewhere. Figure 5.6 shows how warranties can be used to give the eventual building user a right to seek redress from the designers and constructors.

The decision whether or not to reuse existing piles will usually be made after a desk study and preliminary investigations have been completed. It is recommended that a report is prepared setting out the options for the client, explaining the risks and seeking to evaluate the impact in terms of time, cost and quality of the risk materialising (*Section 4.3*). Well-advised professionals will ensure they have their client's informed consent to reuse.

Engineer Designed

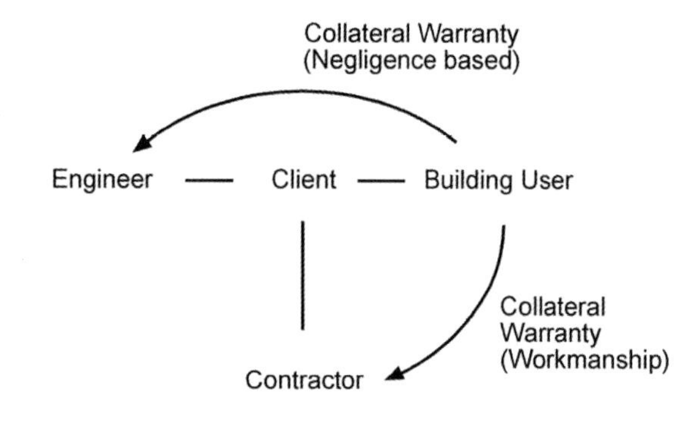

Design / Build

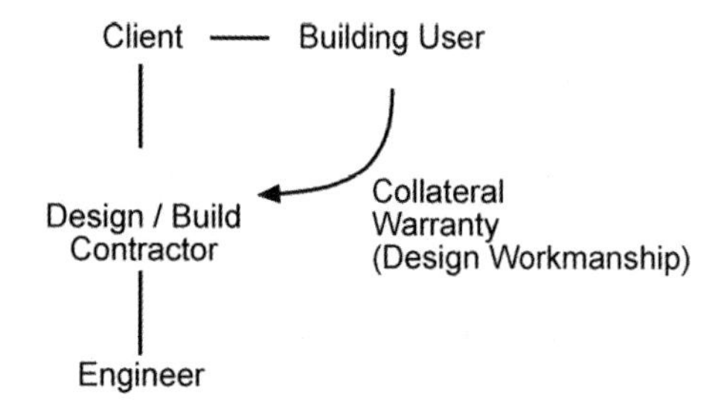

Figure 5.6 Typical UK construction contract structures

A discussion of the respective motivations of the various types of building client (developers, funders and users) is given in *Section 4.2.3.*

A project's commercial viability is often driven by the letting market. The developer client will often aim to sell on, or let on a long lease, the completed building. Tenants/purchasers may have negative perceptions about a building founded on reused piles and this may impact on the client's return. A solution is latent defects insurance (LDI). This covers repair costs (but not necessarily consequential losses), but only following practical completion. It is generally available in the UK and has been adopted for buildings incorporating reused foundations in institutionally funded developments. It applies for a fixed term: typically 12 years. It is similar to the decennial insurance that operates in France and in other countries whose legal systems derive from the French legal system. LDI providers normally require auditing of the design and construction by another firm of consulting engineers; this extra requirement by itself will help to reduce the likelihood of defects.

Where the client is planning to occupy the building for own use, he may be more prepared to consider reused foundations as he may be less concerned about perceived risks impacting on marketability.

5.2.7 Design codes and regulatory approval

Reuse of foundations is not explicitly addressed by current European standards (BS EN 1997-1) and most existing national codes. However, design codes evolve and future revisions of these standards or country-specific National Annexes to the Eurocodes may provide explicit advice on re-engineering of existing foundations. For example, the Building Code of New York City (NYC Buildings 2003) does specifically address the 'use of existing piles at demolished structures' with requirements for assessing the load-bearing capacity for reuse.

The provision of a foundation system incorporating reused foundations may not comply fully with current codes, particularly for materials, so a pragmatic approach is required. As an example, in the UK the proportions of cement needed to resist sulfate attack increase with each evolution of the published advice. Yet, in most circumstances, concrete from a previous era that has survived buried in a ground environment without any significant deterioration should be able to have another design life safely assigned, irrespective of current guidelines for new structures.

To adopt foundation reuse, the client will need to explicitly agree to accept a building that does not fully comply with all aspects of current standards. This acceptance will also need to be agreed with any financial backers, insurers and also regulatory bodies such as any municipal checking office that controls new buildings (such as Building Control Offices/District Surveyor in the UK) or other checking engineer. Despite not being fully compliant with current standards, foundation reuse can be a viable option to support a new building and can be a key to unlocking land for redevelopment that would otherwise be sterilised by the presence of archaeology of buried obstructions.

5.3 Foundation reuse decision process

The first fundamental decision of the foundation strategy for a new redevelopment is the type of foundations: deep or shallow. Normally, the choice between deep and shallow foundations will be dictated by the structural form of the new building (number of storeys, column grid, presence of basement, settlement performance, relative costs) and the particular ground conditions beneath the site. The presence of immovable obstructions in the ground such as old foundations or the presence of archaeology may modify this choice.

Two foundation reuse decision process flow charts have been developed. The first flow chart, shown in Figure 5.7, details the decision process for sites where deep foundations have been chosen as the appropriate foundation solution for the redevelopment and the second, shown in Figure 5.8, details the process for shallow foundations. Although both types of foundation can be reused, there are differences in the approach recommended for deciding if reuse is feasible.

The decision model that is presented in this chapter is a generalised approach suitable for most situations. It is intended as a broad guide and should not be used as a substitute for experienced engineering input to address the particular engineering issues of a project.

5.3.1 Decision flow chart for redevelopments incorporating deep foundations

Where deep foundations have been chosen as the appropriate foundation solution, refer to Figure 5.7 for the foundation decision process.

Presence

Shallow foundations present on the site are unlikely to be suitable for incorporation into the final deep foundation strategy. Existing shallow foundations will therefore form an obstruction to piling and should be removed or cored through as part of the piling works. Installation of any new piled foundations should be carried out to improve the value of the site by maximising opportunities for their future reuse as set out in *Chapter 7*.

Geometric compatibility

The first stage of assessing the potential for reuse is consideration of the geometric compatibility of the existing and proposed foundation locations. Full compatibility will often not be possible, however partial compatibility leading to partial reuse should also be considered. As discussed in *Section 5.2.3*, modifications to the building layout may improve the potential for foundation reuse. However, the implications for the function of the building and the whole life cost of the building should be assessed.

Where there is insufficient compatibility for reuse and it is not feasible to modify the layout of the structure to achieve sufficient compatibility, it will be necessary to construct new foundations. Where existing piles align with the proposed column locations there will be several options for overcoming the location conflict as shown in Figure 5.1, ie:

- remove existing pile and construct new pile at the same location,
- construct new pile(s) offset from the idealised location and construct a large pile cap or transfer structure between the column and new pile(s).

The implications of both of these options should be carefully considered: the effort and time needed to remove the old foundations versus any impairment in the economical functioning of the completed building due to the presence of transfer structures.

Acceptability

If the consideration of compatibility indicates that foundation reuse is feasible, the possibility of foundation reuse should be presented by the design team to the client to assess his attitude towards this foundation option and in particular the different risk and liability associated with it compared with a standard piling contract (see *Section 5.2.6*).

The acceptance of a foundation reuse approach by the client, project funders, insurers and regulatory approval is essential when considering foundation reuse and therefore acceptance 'in principle' should be sought at an early stage of the process. There is often a great temptation for the design team to focus on the technical issues and progress a reuse scheme that has not received any acceptance and by doing so, run the risk of costs and programme over-runs if reuse is found to be unacceptable.

Preliminary assessment

With client approval to consider the potential for foundation reuse, a desk study and preliminary ground investigation should be undertaken to determine the ground conditions and to assess the capacity, durability and quality of the existing foundations as described in *Chapter 6*. Where the desk study and investigation identifies that the existing records are poor or where there is an indication of failure of the old foundations from the performance of the previous structure, foundation reuse may not be feasible. It may be possible to proceed with reuse by downgrading the design capacity of the existing pile, but where this is not feasible new foundations will be required. At the end of the desk study and investigation, a report should be prepared setting out the options for the client and explaining the relative risks.

Foundation option selection

If foundation reuse is proven to be feasible through all these decision stages, it is appropriate to assess the benefits over other foundation strategies such as the construction of new deep foundations. Several assessment methods and tools are available depending on the priorities of the client. The assessment criteria typically include:

- construction cost and programme,
- whole life building cost,
- environmental impact or sustainability.

Further discussion of these assessment criteria for foundation reuse is presented in *Section 5.4*.

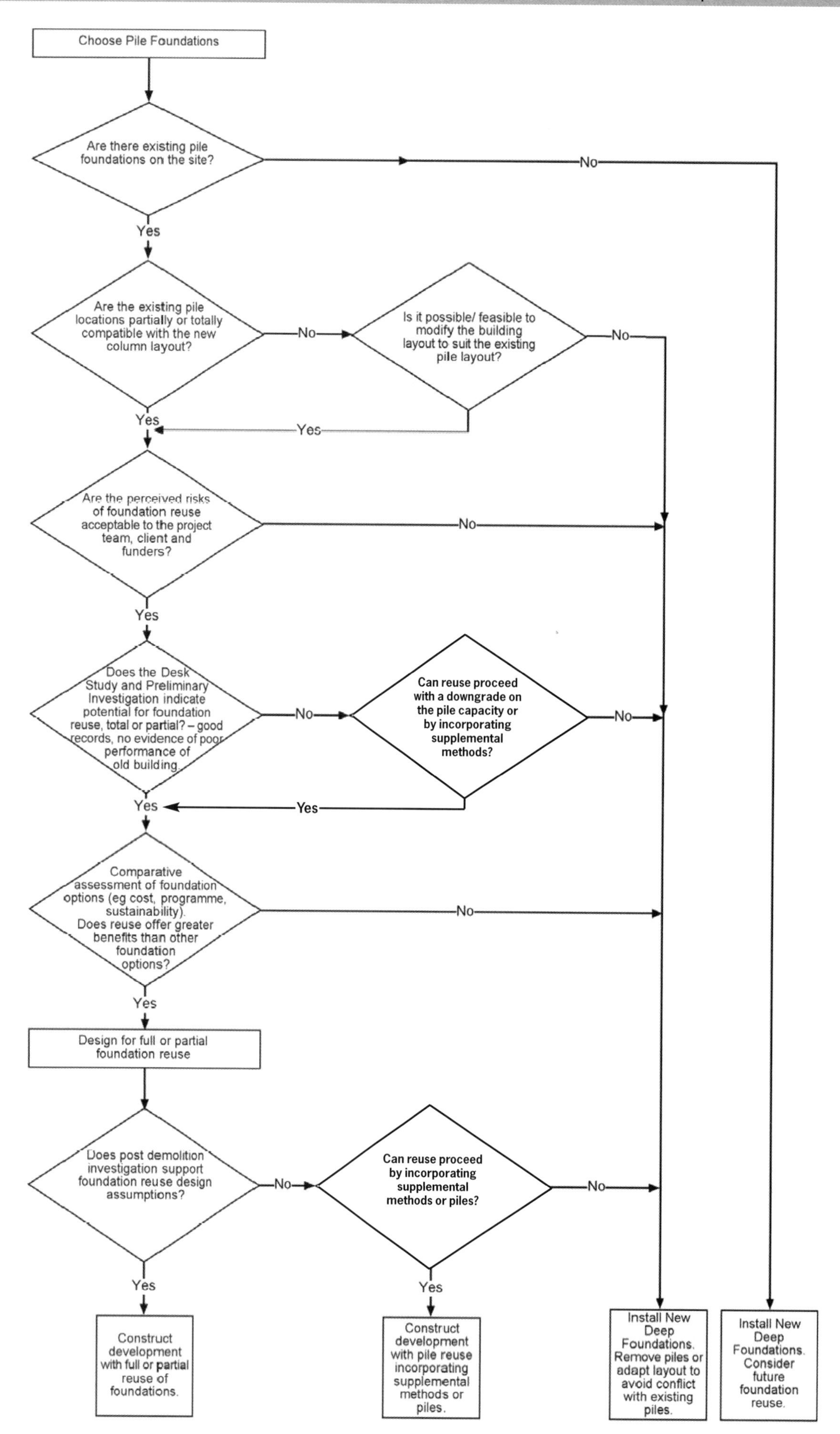

Figure 5.7 Foundation reuse decision process: deep foundations

Design
Where the assessment process results in the acceptance of foundation reuse, the design for full or partial reuse can proceed as outlined in *Section 3.4.5* and *Chapter 6*.

Final assessment
Verification of the records identified in the desk study through intrusive investigations will usually be required (*Chapter 6*). Some of this investigation work is likely to take place when demolition of the existing building is in progress or complete, allowing access to the foundations. If the available information cannot be verified and investigation does not support the foundation reuse design assumptions, the foundations cannot be reliably reused and new foundations will be required.

5.3.2 Decision flow chart for redevelopment incorporating shallow foundations

Where shallow foundations have been chosen as the appropriate foundation solution, refer to Figure 5.8 for the foundation decision process.

Presence
If there are no existing foundations on the site, foundation reuse will not be possible and new shallow foundations will be required. Where existing deep foundations are present on site, there is potential for them to be suitable for reuse. For example, pile reuse to assist new shallow foundations will be a solution where a low-rise structure is being built on a site previously occupied by a heavier structure, for instance when an old power plant is being redeveloped into light industrial use or a warehouse is being redeveloped into housing. Another example is where old piled foundations are treated like ground improvement to allow a raft to be built over the top where there would otherwise have been concerns about excessive settlements. Where existing shallow foundations are present on a site, there is potential for them to be suitable for reuse.

Preliminary assessment
The feasibility of reusing the foundations will depend on the balance between the difficulties in removing them and the capacity and performance benefits from incorporating them into the new foundations. For shallow foundation solutions, ie low-rise buildings replacing older low-rise structures, removal of the previous foundations and construction of new shallow foundations may be relatively inexpensive and the most efficient solution. However, the cost-effectiveness of removing existing shallow foundations may be affected by their size, the presence of archaeological deposits or by future changes in waste classification or landfill costs.

On occasions, the continued presence of the old foundations will not be helpful to the new foundations, and steps will need to be taken to minimise their interaction. For instance, where the new building just partially straddles the edge of the old foundations, steps may need to be taken to prevent damaging differential movements between the two parts of the building and this may dominate the foundation design. For this situation, more economical and robust foundations may be achieved if it is possible to move the structure wholly to overlie the old foundations, or to move the structure wholly away from them.

Where the existing shallow foundations are difficult to remove or are helpful for improving the performance of the new foundations, the possibility of foundation reuse should be presented by the design team to the client to assess their attitude towards this foundation option.

The remainder of the decision process for shallow foundations is similar to deep foundations as described in *Section 5.3.1*. However, for shallow foundations the degree of testing and verification may be less extensive than for deep foundations, as they are normally more robust and simpler to verify.

5.4 Risk management and foundation choice assessment methods

Traditionally, it has been assumed that the cost of a building was limited to its construction, and that the interest of the client was best served by delivering the cheapest constructed solution. This is now changing and it has been recognised that the engineering solution with the least cost may not provide the best value solution for the client. The majority of developers are interested in making an income from the development, which is linked to when the building will be available for use and its maintenance during the design life. Therefore, it is maximising the combined benefit of construction costs, programme and maintenance requirements plus any additional client criteria that will provide the best value to the client.

Risk management is a key process in successful delivery of construction projects. Risks to the successful completion of the project are identified and assessed to consider their impact on the project. In the case of foundation reuse, risk management can also be used to assess any issues of concern and thereby dispel misconceptions as to the perceived risk associated with this solution. Risk management can also be used to consider the benefits of different options under consideration (eg foundation reuse) and therefore to inform the decision process.

> *Note:* The process of project risk management should not replace the process of identification, elimination and management of risks to health and safety.

Project risk management strategies are well-documented [ICE/IoA (1998), ICE/DETR (2001), CIRIA (1996), APM (1997), Project Management Institute (2000)]. Risk management should run throughout the life of the projects and may include the following items.

- *Risk management plan*. This describes how risk will be managed on the project. The plan is set up at the outset of the project and is updated throughout the project.
- *Risk register*. This provides a summary of the risks that may affect a project and the actions that are proposed to

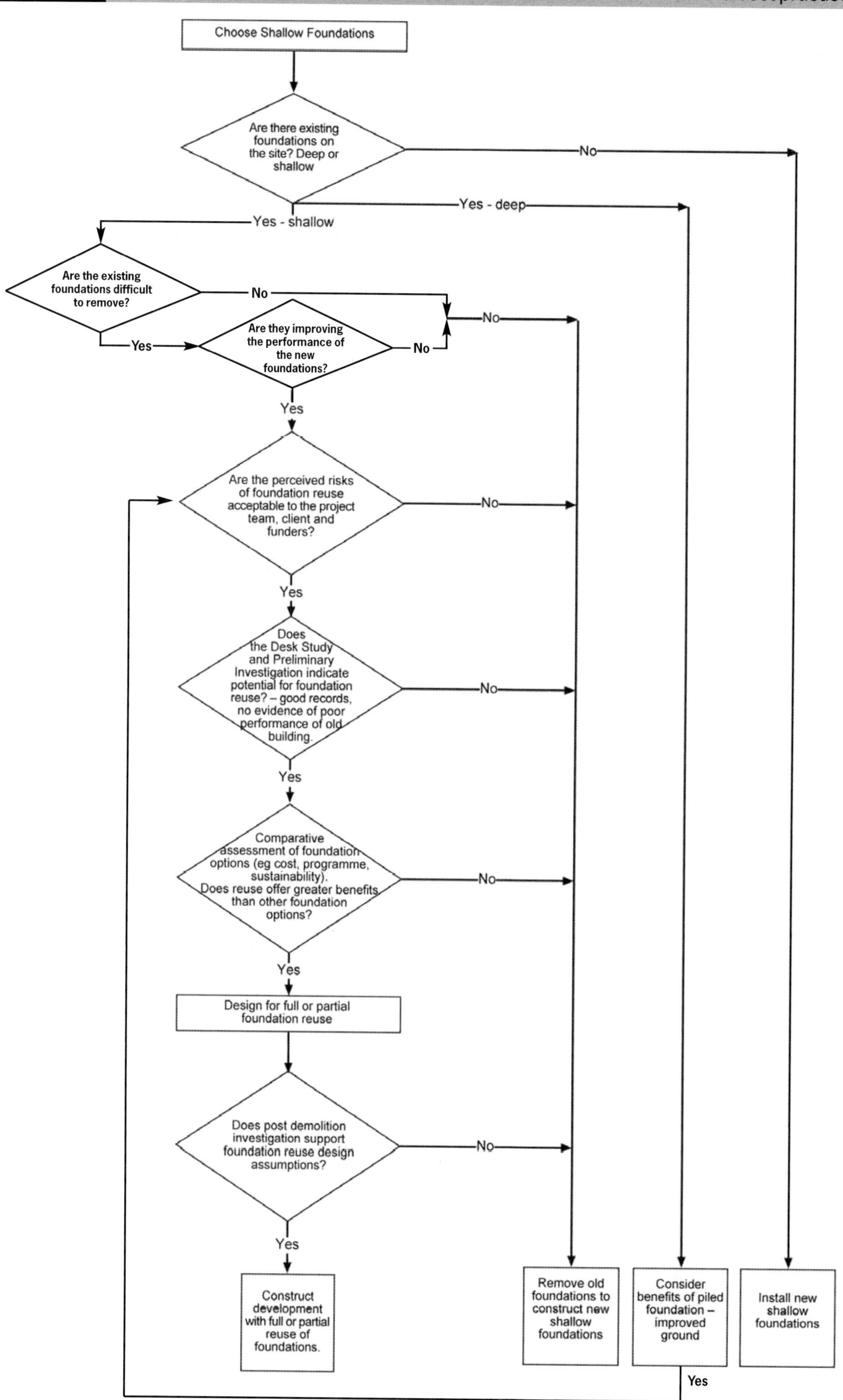

Figure 5.8 Foundation reuse decision process: shallow foundations

deal with them. The risk register is set up at the outset of the project and is updated throughout the project.

- *Qualitative or semi-quantitative risk assessment.* This is a method for assessing the importance of a risk without necessarily being able to quantify it. It is usually used in the early stages of a project when not enough is known about a risk to be able to assess it quantitatively.
- *Quantitative risk assessment.* This is a method for quantifying risks and assessing the possible range of outcomes for the project (usually in terms of cost and/or time). It is usually carried out at the beginning of design and updated at key stages in the project as more detailed information is available.

Probabilistic approaches [Harr (1987), Stille et al (2003), Norrman (2004) and Isaksson (2002)] and in particular the use of probabilistic risk analysis tools to undertake a quantitative risk analysis is increasing in the civil engineering industry. These analyses provide a way of quantifying the potential financial exposure that a client may have when undertaking a particular project.

Probabilistic risk analysis tools such as @Risk (© Palisade Corporation 2004) consider the probability and variability of different items (eg costs, programme duration or energy inputs) to determine the probability of different cost, programme or energy outcomes. It can be used to assess the viability of a project or can be used to compare different alternatives such as foundation reuse and complete foundation replacement. An example of a probabilistic risk analysis is presented for an idealised case history in Appendix B. Many probabilistic analysis tools can be used to identify clearly the largest risk items for the project which will help to inform the project team where best to focus on risk mitigation as part of a risk management process.

5.4.1 Construction cost and programme

The assessment of redevelopment costs is often used as an assessment criterion when considering the feasibility of redeveloping a site or when assessing the benefits of different alternatives. An assessment of redevelopment costs, including foundations, will need to include:

- direct costs (eg material costs),
- risk costs (eg additional foundation costs, delayed use of building if post-demolition investigation reveals that foundation reuse is unsuitable),
- benefits (eg earlier occupation of the building, future redevelopment options).

Direct costs include the known construction costs. However, for most known costs, there will be some degree of variability such as the unit price of steel or other materials.

In most cases, foundation direct costs are low in comparison to the main superstructure direct costs. However, foundations is the area that can make even the most experienced client cautious due to the relatively high incidence of construction cost and programme over-runs resulting from ground-related problems. Chapman & Marcetteau (2004) have demonstrated that for a typical UK building project, the site investigation costs are usually just 2% of the substructure costs or 0.1% of the overall building cost despite ground-related problems typically accounting for about one-third to one-half of construction programme over-runs.

Risk costs represent uncertain events, ie events that may or may not occur. For example, there is a risk that the post-demolition investigation will identify that the existing foundations are not suitable for reuse which will result in additional design and construction costs as well as a delay to the construction programme. These costs are considered as risk costs as they will only occur if the investigation invalidates the design assumptions. Frequently, the risk costs can change significantly as the particular circumstances associated with the risk are often not precise. In the previous example, the financial implications will be larger if all of the foundations are identified to be unsuitable rather than only a few of the foundations. The best way to reduce risk on a project is to identify the potential hazards that need to be eliminated, reduced or mitigated against. However, a certain amount of risk may still apply which the client may wish to quantify.

Benefits can be considered as negative costs or income. A key benefit for redevelopment projects is usually early occupation of the building because rental income will ensue.

5.4.1.1 Direct costs

Design costs

Both installing new foundations and reusing old foundations will incur costs by a design team, including structural and geotechnical engineers.

When full or partial foundation reuse is to be considered the design cost should make allowance for the more thorough desk study as well as additional design and assessment of intrusive investigations as described in *Chapter 6*.

For full or partial foundation reuse, the design costs will be dependent on the compatibility of the new column loads and the existing foundations. Where these are not compatible, transfer structures will need to be designed to compensate for the incompatibility. This will also affect construction costs.

Partial replacement is more likely than full replacement due to issues such as compatibility. Due to differing settlement performance between the new and existing foundations, additional assessment of settlement performance will be required.

The design costs associated with complete replacement of the existing foundations will also be influenced by the compatibility of location of the new building loads and any existing obstruction or foundations in the ground. Where there is compatibility between the building loads and obstructions in the ground and these cannot be removed, the new foundations will have to be installed around the existing obstructions and transfer structures or non-standard piling layouts will need to be designed.

Investigation costs
The scope of intrusive investigation will depend on the quality and quantity of records of the existing foundation identified in the desk study.

Where foundation reuse is chosen and undertaken in a responsible manner as described in this *Handbook*, intrusive investigations will be required to confirm the desk study information, identify unknown aspects of the foundations, such as pile dimensions, and to identify and assess any material deterioration that may have occurred.

Where new foundations are to be installed on a congested site, additional investigation may also be required to identify available locations for the new foundations.

Construction costs
The cost of a normal full replacement piling contract is likely to be the easiest cost to quantify. The cost will include material, labour and plant, and will depend on the duration of the construction programme. If the existing foundations are to be reused with no requirements for supplemental foundations, there will be no new pile construction cost.

If new piles are to be installed either as part of a partial or complete replacement solution, there will be costs associated with the new foundations. Depending on the compatibility between the new building loads and the existing foundations, transfer structures may be required which will have associated direct costs.

Insurance costs
Additional insurance such as defects liability insurance or decennial insurance may be appropriate for foundation reuse schemes as identified in *Section 5.2.6*. The premiums associated with this additional insurance will increase the direct costs on a project. If the insurer imposes an exclusion or limit on the insured loss, the risk may need to be revisited to make sure that the client or owner are able to accept the exclusion.

5.4.1.2 Risk costs

Potential risks to the financial success of the project should be identified as part of the risk management process described previously in this chapter. Risk items can vary enormously depending on the particularities of the site.

One of the risks associated with adopting a foundation reuse strategy is the late identification of the unsuitability of the foundations for reuse. Often, confirmation of the suitability for foundation reuse can only be obtained during or post demolition investigation of the existing structure. If the investigation reveals that foundation reuse is not suitable, new foundations will be required. This can have a significant effect on the project completion as further substructure and possibly superstructure design will now be required, followed by procurement and construction of the new deep foundations. Typically, foundation construction is on the critical programme path and therefore these delays will have a direct influence on the overall construction programme.

A prolonged programme can have a significant effect on the viability of the project as it will:

- result in additional design and construction costs,
- delay the occupation of the building and hence the generation of income for the client,
- impair the client's ability to repay any monies borrowed to cover construction costs.

5.4.1.3 Financial benefits

Benefits can be seen as negative costs (eg the early occupation of the building will lead to income in the form of rent which may also allow the developer to commence repayment of any loans used for the redevelopment. While much of this *Handbook* is devoted to countering the perception of increased risks arising from foundation reuse, the solution can also bring about several benefits. These potentially include:

- a more rapid construction programme resulting in earlier occupation,
- reduction in foundation construction costs,
- less damage to archaeology, and possibly less need for archaeological intervention on the site,
- improved sustainability, eg reduced use of raw materials for the new foundations and less energy to install them.

5.4.2 Whole life costs

Whole life (or life cycle) costing (WLC) can be used as a tool to assess the total cost performance of a building throughout its operational life.

A simple definition of WLC is:

> 'the systematic consideration of all relevant costs and revenues associated with the ownership of an asset.'

An example of WLC is given by Chapman & Marcetteau (2004) who show that for a typical office building, the value of economic endeavour within it over its life will be of the order of 30 times the building cost and over the building life the operating costs will be of a similar magnitude to the building cost.

5.4.2.1 Uses of WLC

WLC is a tool to assist in making decisions between different options with different cash flows over a period of time. In this respect, it is a form of investment analysis. As future costs and revenues are discounted to present day values, the earlier a cost occurs, the greater impact it will have. WLC is aimed at facilitating choices where there are alternative means of achieving the client's objectives.

WLC is relevant when considering whole estates, individual buildings or structures and when comparing alternative investment scenarios and associated risks such as:

- retain and refurbish or demolish and redevelop,
- alternative designs (eg foundation reuse or new foundations).

It is particularly used to justify whether an alternative with a higher capital cost is justified and can take into account other advantages associated with various options, such as planning issues, construction timescale, etc. Most benefit will be obtained if WLC is taken into account at the earliest stages of design, and in setting initial budgets since budgets based on previous experience will reflect procurement on the basis of lowest capital costs.

WLC can also be carried out as a risk management exercise (eg in justifying an increased site investigation that may reduce construction costs and risks of delays). WLC calculations are often dominated by in-service costs so the most critical elements may be the potential for delays (and hence lower rents) or any compromise that reduces floor area or efficiency.

5.4.2.2 Basic steps in WLC

Basic steps in WLC are listed below.

- Identify capital and operational costs and incomes.
- Identify when they are likely to occur: for each cost, there should be an associated time profile of when the cost occurs (or recurs).
- Identify upper and lower bounds for these costs to illustrate risk.
- Use 'discounted cash flow' analysis to bring the costs back to a common basis: items should normally be entered into the analysis at the current cost and a 'real' (excluding inflation) discount rate applied. Normally, this will be done on a commercial spreadsheet package which includes equations for discounted cash flow.
- Undertake sensitivity analysis of the variables such as the discount rate, the study period, the predicted design lives of components, etc.

5.4.2.3 WLC in relation to foundation reuse

In the context of foundation reuse, the costs associated with different foundation options should be considered for the following building stages:

- design,
- demolition,
- construction,
- operation, repair and maintenance,
- refurbishment, or demolition and replacement.

In each case, different reuse options may impact on direct costs, on programme or future options such as refurbishment, or on demolition and replacement. It should be noted that data on costs may not be readily available which will limit the effectiveness of the WLC process.

The WLC process can also be extended to take account of opportunities or risks arising from the options selected. An example might be, if a foundation system was selected that would make future development of a heavier building more difficult, it may be less favoured and may devalue an asset compared with one that leaves this option open. Foundation reuse is intrinsically linked with refurbishment options. Medium or major refurbishments may well involve changes to the structural form of the building and hence the loading or loading patterns on the foundations. This will necessitate an assessment of whether the old foundations are adequate for the purpose. Reuse issues could be integrated into a WLC hypothesis for the future usage of the building which would,

Box 5.1 Whole life costing issues relevant to foundation reuse only: redevelopment of a commercial site

Purchase

- Buy land

Design

- Resolve planning issues, including archaeology
- Source old design/construction records and verify
- Site investigation for normal foundation design process
- Site investigation for foundation reuse
- Site investigation for avoidance of existing foundations and obstructions
- Site investigation for removal of existing foundations and obstructions
- Make records
- Foundation design for new foundations
- Foundation design for new foundations to replace those removed
- Foundation design for old foundations
- Foundation design for mixed scheme foundations
- Satisfy relevant regulatory authorities
- Specifications for construction/monitoring
- Other substructure design
- Transfer structure design
- Superstructure design

Demolition

- Install building monitoring/monitor/analyse/report
- Carry out load takedown/analyse/report
- Demolish superstructure
- Remove old foundations/substructure

Construction

- Construction of new foundation elements
- Other substructure elements
- Archaeological works/monitoring
- Insurance
- Safety
- Upgrading of old foundations
- Testing/monitoring
- Superstructure
- Fit-out and finishes
- Close out records

In service

- Normal in-service energy, maintenance
- Compromised layout: possible increased energy use and maintenance over period
- Rent over period
- Compromised layout: reduced rent over period
- Business costs
- Insurance
- Monitoring and keeping records

Future

- Redevelopment costs
- Redevelopment options
- Residual building value
- Residual land value

over the life of the asset, involve interim refurbishment as well as complete redevelopment.

Box 5.1 outlines the issues relating to foundation reuse that may affect the costs or income associated with construction, operation and decommissioning of a structure. An example is given in Appendix C.

5.4.3 Environmental impacts

One benefit of partial or total foundation reuse is that it is likely to have a lower environmental impact than the alternative foundation options, namely removal and replacement of piles or installation of a new foundation system as shown in Figure 5.9.

The removal of less material from site during demolition and the installation of fewer piles have benefits in:

- less noise, machinery time and fewer lorry movements during demolition,
- less machinery time required during construction,
- less concrete used and spoil produced.

This not only adds to the sustainability of a redevelopment project, but can have measurable cost benefits.

To assess the environmental impacts of products, several environmental management techniques are available. One appropriate technique is life-cycle assessment (LCA). This method does not involve economic or social aspects, nevertheless measurable cost benefits may be linked to a sustainable project. LCA is governed by European Standards, EN ISO 14040 *et seqq*. It takes into account all stages of life of a product system and their impact on the environment by compiling and quantifying the materials and energy that enter or leave the product system (life-cycle inventory analysis). Subsequently, the impact on the environment is assessed (life-cycle impact assessment). Some emissions (eg noise) cannot be evaluated by this method as these can only be measured in a qualitative way.

Consideration of overall energy inputs or embodied energy can be used as an indicator of the environmental impact or sustainability of a process or product. A simple definition of embodied energy (EE) is:

> 'the energy required to win the raw materials required, create materials and products and transport them to their point of use'.

Consideration of whole life EE input can also be used as a measurement criterion to compare the benefits of alternative options such as the reuse of foundations compared with installation of a new foundations system through congested ground. For foundations, the embodied energy will include not only the energy embodied in the foundation materials but also the energy required to install them (which can be measured through the embodied energy of fuel required for the plant).

5.4.3.1 Assessment of EE

The process of assessing EE/sources of EE relies on knowledge of energy usage associated with different processes which is often not readily available. However, the assessment of embodied energy will become easier in the future when standard values become accepted for different processes. The following steps should be considered for an EE assessment.

- Energy in required construction materials and products.
- Where recycled materials are incorporated in a product their embodied energy is calculated from the point at which the materials would have been considered as waste.
- Energy used in the construction process: the energy is calculated as that used at source, as opposed to that which is delivered as electricity. This is termed primary energy.

Very broadly, the more processing an item undergoes during its production, the higher the EE is likely to be, and therefore the greater the impact it has on the environment.

5.4.3.2 Assessment of CO_2

Embodied carbon dioxide (CO_2) is a more direct measure of environmental impact, namely that of global warming where

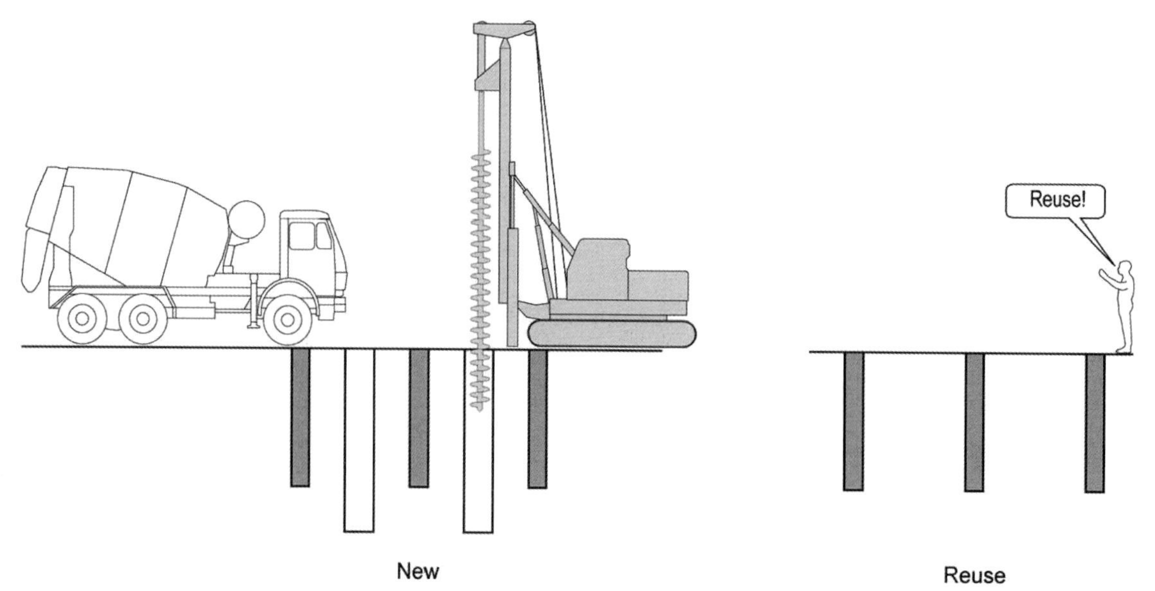

Figure 5.9 Environmental savings through foundation reuse

global warming potential is calculated as kg CO_2 equivalent. It is based on the UK power generation mix and the efficiency of materials production processes.

Embodied energy or CO_2 values should not be used directly to compare one material against another and the comparison should always be made between components that perform the same task. The embodied energy of steel is much higher per tonne than concrete which indicates a preference for concrete if incorrectly compared. However, if a comparable specification steel-framed and concrete-framed building are evaluated by calculating EE per m^2 floor area, it will be found that the EE or CO_2 are similar and therefore consideration of EE or CO_2 will not show a strong preference for either solution.

5.4.3.3 EE in relation to foundation reuse

A simple comparison is made in Appendix C of the environmental impacts during the construction process of two of these options, based on a case study. The analysis has been greatly simplified by using energy input into raw materials and construction processes only, as a quantifiable measure of environmental impact. Other impacts, such as waste and pollution are described but not quantified. In the case study analysed, reuse of the foundations where possible would have resulted in less than half of the energy required and would have significantly reduced noise, dust and other pollution associated with demolition, transport and construction activities.

5.5 Key points

The ability to reuse foundations will depend on the technical complexities of each particular site. For foundation reuse to be successful, two basic technical principles must apply.

- There must be geometrical compatibility between the locations of the applied loads and the existing foundations, which must have sufficient capacity to carry the new loads and have settlement characteristics appropriate to the new structure (referred to as settlement compatibility in *Chapter 6*).
- Sufficient verification must be carried out so that the old foundations are shown to be as reliable as new ones.

Consideration of foundation reuse will also be influenced by the particular values of the client and project team as well as their attitude to uncertainty. For the decision to reuse to be acceptable, the following should apply.

- All members of the design and construction team should embrace a foundation strategy involving reuse.
- The client is prepared to consider such a strategy, and is able to obtain any suitable insurance products required to carry some of the risks.
- The approach also proves acceptable to the necessary authorities.

Key issues relating to foundation reuse have been identified and used to develop a foundation reuse decision process with the aim of enabling owners, funders, insurers, regulators and designers of proposed redevelopment schemes to evaluate the feasibility of reusing existing foundations.

Like any other construction option considered, a foundation reuse strategy must offer advantages over other foundation strategies, in terms of the criteria that are important to the project stakeholders. Typical criteria or indicators include:

- construction cost and programme,
- whole life cost,
- environmental impact or sustainability.

5.6 References

Association of Project Management (APM). *Project risk analysis and management guide*, 2nd edition. High Wycombe, APM, 1997

British Standards Institution. BS EN ISO 14040: 1997 *Environmental management. Life cycle assessment. Principles and framework*

British Standards Institution. BS EN 1997-1: 2004 *Eurocode 7. Geotechnical design. General rules*

Chapman T & Marcetteau A. Achieving economy and reliability in piled foundation design for a building project. *The Structural Engineer* 2004: **82** (11)

European Union (Employment and social affairs). *Council Directive 92/57/EEC of the European Parliament and of the Council of 24 June 1992 on the implementation of minimum safety and health requirements at temporary or mobile construction sites.* Luxembourg, Office for Official Publications of the European Communities, 2000

Godfrey PS. *Control of risk: a guide to the systematic management of risk from construction.* Special Publication 125 (SP 125). London, CIRIA, 1996

Gold CA & Martin AJ (BCA/BSRIA). *Refurbishment of concrete buildings: structural and services options.* Guidance Note GN8/99. Bracknell, BSRIA, 1999

Harr ME. *Reliability-based design in civil engineering.* Mineola, New York, Dover Publications Inc. 1987

HMSO. *The Construction (Design and Management) Regulations 1994.* Statutory Instrument 1994 No. 3140. London, The Stationery Office, 1994

Institution of Civil Engineers (ICE)/Faculty of Actuaries/Institute of Actuaries (IoA). Risk analysis and management for projects (RAMP). London, Thomas Telford, 1998. See also www.ramprisk.com

Institution of Civil Engineers (ICE). *Managing geotechnical risk: improving productivity in UK building and construction.* London, Thomas Telford, 2001

Isaksson T. *Model for estimation of time and cost based on risk evaluation applied on tunnel projects.* PhD Thesis, Royal Institute of Technology, Stockholm, 2002

Norrman J. *On Bayesian decision analysis for evaluating alternative actions at contaminated sites.* PhD Thesis, Chalmers University of Technology, Goteborg, 2004

NYC Buildings. *Building Code of the City of New York.* New York City Department of Buildings, New York, 2003
See www.nyc.gov/html/dob/html/reference/code_internet.shtml

Palisade Corporation. @Risk software. 2004
See www.palisade-europe.com

Project Management Institute (PMI) A guide to the project management body of knowledge (PMBOK® guide). Danver, Massachusetts, PMI, 2000

Stille H, Andersson J & Olsson L. *Information based design in rock engineering.* Report 61. Stockholm, Swedish Rock Engineering Research, 2003

6 Investigation, assessment and design of reused foundations

6.1 Introduction

This chapter introduces the technical and engineering issues that need to be addressed in the investigation, assessment and design of foundations for reuse.

Reuse engineering readily splits into assessment and design. The best foundation provision is produced when the structural engineer and the geotechnical specialist work closely together in an iterative process. Figure 6.1 shows a flow chart for the investigation and assessment of foundations for reuse and emphasises how much of a central tenet continuous assessment is to the reuse process. Iterative processes are fundamental to all reuse activities compared with 'greenfield' sites; reuse investigators must be prepared to assess this feature continually.

Practices and concepts employed in foundation provision for the 'greenfield' case are applicable. Fundamental principles cannot be ignored:

- a comprehensive knowledge of the ground (the foundation medium) is necessary and can be obtained by a comprehensive desk study and walkover survey backed up by physical investigations to fill in the gaps in knowledge.

These activities need to be adjusted from that needed for 'greenfield' sites to investigate both the ground and the existing foundations that have the potential for reuse.

The condition of the existing foundations can be assessed using non-destructive testing (NDT) for durability and integrity of components. The load capacity performance can be assessed from measurements during demolition and/or load testing to prove they are acceptable for the new superstructure. Assessment of load-settlement behaviour must ensure that the risk of brittle failure is kept to an acceptably low level.

The above processes with the continual viability assessment interacting with the design will show the potential reserves of capacity in the existing foundations to be reused from time-related effects and unused capacity in structural elements. The design will allow the determination of the requirement, if necessary, for supplemental methods, whether global or local improvement techniques. In this case, the consideration of strain compatibility is paramount to ensure the old and new foundations work in harmony.

Each stage of the reuse process must be thoroughly verified. Each assessment and design stage must be evaluated objectively within a risk-aware framework. Verification is the use of a prescribed observational method to ensure continuing compatibility between old and new foundation elements.

The final part of this chapter includes three case studies that demonstrate the points discussed.

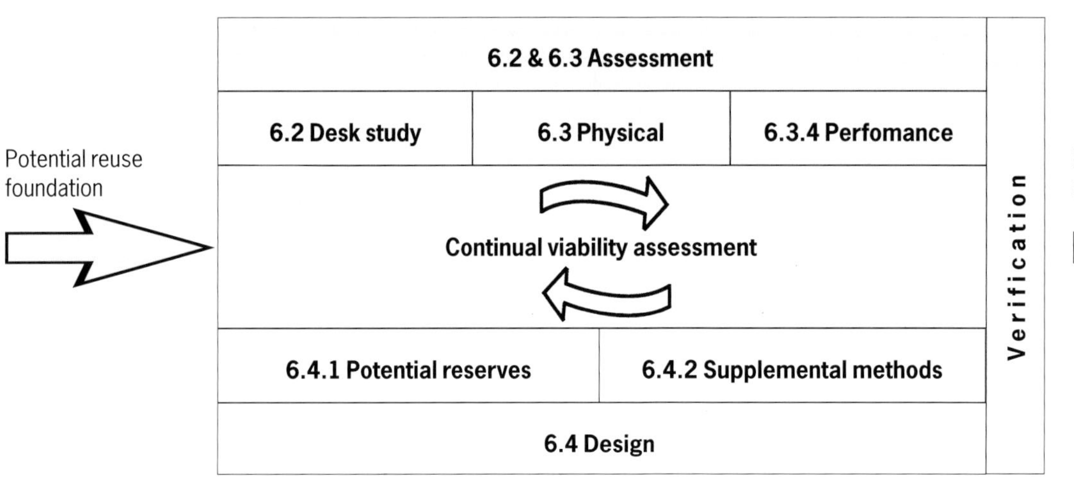

Figure 6.1 Flow chart for the investigation and assessment of foundations for reuse

6.2 Desk study for reuse

6.2.1 Importance of desk study and walkover survey

The desk study (see *BRE Digest 318*) is the most cost-effective way of gathering information about the development site and the local environs. It is important to collect and collate all available information before undertaking the walkover survey (see *BRE Digest 348*). The desk study will reveal gaps in the information available, some of which can be found during the walkover survey. The walkover survey will also supply information that will confirm or augment the desk study data and additonally supply information about the local environs to the site. The information gathered in the desk study and the walkover survey will be used to design the ground investigation and will include elements to assess the existing foundations for reuse or removal.

In this section, only those points in the desk study and walkover survey relevant to the reuse of foundations will be discussed.

6.2.2 Sources of information

Figure 6.2 shows the knowledge needed for successful investigation into the reuse of foundations but it is merely an extension of that needed for a normal site investigation. It is crucial to collect all available information about the existing foundation and ground. Possible sources of information are listed in Box 6.1 and described in more detail in the sections that follow.

Box 6.1 Sources of information for the desk study

- Current owner/occupier
- Previous owners/occupiers
- Statutory regulations records
- Land condition record
- Local/municipal authority (planning and building control)
- Archaeological agency
- Infrastructure companies (road, rail, water, etc.)
- Foundation contractor
- Foundation designer
- Other historical information from local history/museum organisations

Current owner/occupier

The current owner, depending on the time of ownership/occupation, could have records that include details of the building or at the very least the age of the building and the architects, engineers and contractors who were involved in its construction. These records will provide leads to other sources of information.

Previous owners/occupiers

Previous owners can be found through local authority/municipal records or from the current owners and, again, can provide both information and leads to other sources of information.

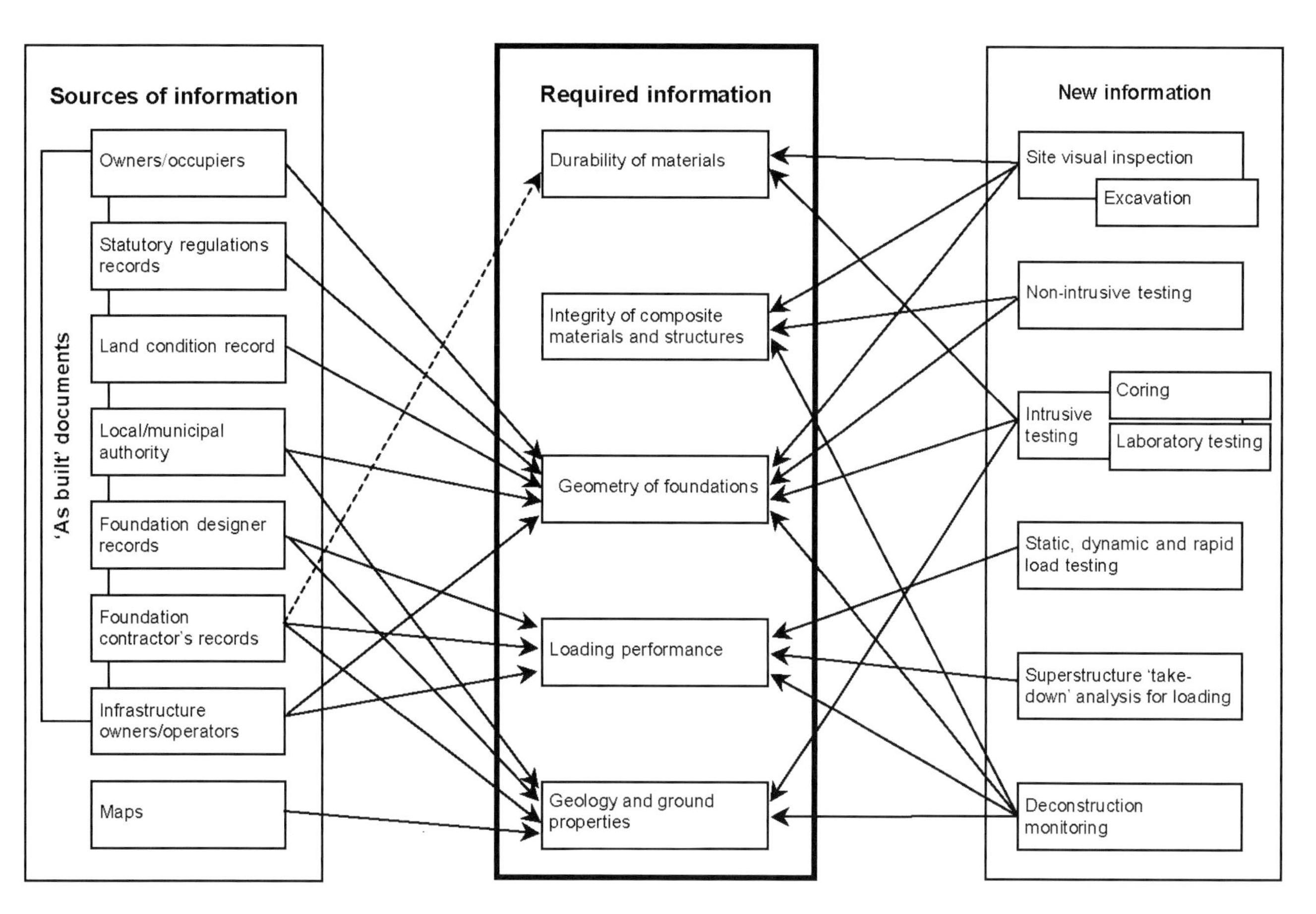

Figure 6.2 Knowledge needed for successful investigation into the reuse of foundations

Statutory regulations records
Many countries now demand that records should be made during construction about construction activities and that these records are kept by the owner or occupier of the building. These records should include data on the construction of the foundations.

Land condition record
In many countries, the national or regional authorities have a registry where all the land is registered so present and past owners are listed and, in some cases, the land use and locations of developments.

Local/municipal authority (planning and building control)
The local authority/municipality will have records that contain valuable information about all the buildings in their area of authority. In general, the local authority/municipality will have granted permission to construct the structure and during that process will have been furnished with plans and details of foundations. Such plans will include the proposed foundations but care should be taken as often the planned foundation is changed by the foundation contractor for a more economical solution or the foundation is changed during construction and 'as built' records may not have been supplied to the planning/building control authority. It is the 'as built' records that are most important.

Archaeological agency
Many cities and towns are known to be located on the site of ancient settlements so archaeological remains will be present. The local authority or municipal offices will know the archaeological agency that has most experience of the local area. They will have excavated archaeological remains in the area and will probably have a map of the ancient settlements and knowledge of the most recent finds. Their guidance on the likely archaeological significance of the site will help not only as a driver for reuse to preserve the archaeology but also in project planning as local or national regulations will require the excavation of the site to determine the extent and significance of the remains.

Infrastructure companies (road, rail, water, etc.)
Infrastructure companies who own/operate/maintain tunnels, roads and other such structures that may be adjacent to or run through a site will have statutory rights and provisions over activities on the site but will also have records as to activities in the past (eg when tunnels were driven or trenches excavated and what was found, and any problems that occurred and were solved. This information is particularly useful where work was completed after the existing structure was built.

Foundation contractor
Foundation contractors keep records of their work for about 12 years for insurance purposes but tend to destroy older records. However, they may have information about the typical foundation types that were installed in the area and at the time of construction of the existing building. This information will help in determining the type of the existing foundation and possibly the reinforcing bar design, and may include copies of the ground investigation reports, pile test reports and reports of instrumentation data.

Foundation designer
Foundation designers tend to keep records longer because of possible follow-on work from a good client and are more likely to have design drawings and calculations, site investigation reports and planning application reports. Again, as for the foundation contractor they will have information about the typical foundation types used in the area and location of the site but will also include more detailed information about the design loads, expected settlements and reinforcing steel layout.

Other historical information
Local history societies or local libraries often keep local items of history, including local newspaper archives that could contain dates of construction or pictures showing the work. This will reveal the age and maybe much more information. The local history society, local library or municipal archive will also reveal information about the site and surrounding area's archaeological background.

6.2.3 Walkover survey

The walkover survey should be undertaken after much of the desk study information has been completed. It can be used to both confirm and augment the desk study information, filling in any gaps in the information so far gathered. After the walkover survey, the completion of the rest of the desk study and planning of the intrusive site investigation can be focussed on filling in additional vital/required missing information.

The walkover survey will comprise these five elements:

- collecting information not found in the desk study,
- confirming information found in the desk study,
- visual inspection of the building,
- visual inspection of the outside of surrounding buildings (and inside where possible),
- acquiring local knowledge.

The walkover survey will start with a list, based on the desk study, of items that are missing and information that needs confirmation. This could, in the worst case of minimal data about the foundations, be the spacing of the columns and whether pile caps can be identified and some indications about the age of the building through to the construction method, cladding, internal fittings, etc. For a building that has been well-documented, the exercise will be mainly to confirm the location of supporting columns, retaining walls, etc.

The visual inspection of the building will be made to identify any structural distress (identified by cracks in structural members, misalignment of cladding panels, signs of repairs, etc.). Any structural distress will show a possible change in distribution of loading that will need to be recognised. More importantly, the cause of the redistribution of load will need to be established. The most critical is if the distress is due to foundation settlement, as this will need further investigation before foundation reuse can be considered. Other factors that can cause building foundations to settle need to be checked (eg subsurface structures and tunnels constructed after the building was constructed).

The visual inspection of the buildings in the immediate vicinity of the proposed redevelopment can indicate likely foundation problems. As for the inspection of the building that is to be redeveloped structural distress can help to identify problems with foundations or other influences such as subsurface structures.

Local knowledge can be acquired from documentation and newspaper cuttings held in the local library and from anecdotal evidence provided by local inhabitants and history societies.

6.2.4 Required information

Information relevant to a reuse of foundations site will include the following.

Site

- Location
- Local topography
- Current and previous use
- Current and previous use of adjacent plots

Ground

- Geology, recent (last 100 years) history
- Ground properties
- Groundwater

Foundations

- Design (piled raft, suspended basement slab, independent or integral retaining walls)
- Type (piles, ground slab, deep footings, ground treatment)
- Design loads (per pile, per pile group, transfer loads, column loads)
- Geometry (pile groups: number of piles, spacing of groups, retaining walls)
- Date of installation
- Pile/foundation load tests
- Foundation construction method (piles: bored, driven, cast-in-situ, continuous flight auger; reinforcing: number, depth, spacer links, top-down construction of below-ground floors, strutted retaining walls)
- Design specification of materials (concrete design mix, steel material specification)
- Current condition (integrity of component parts, condition of component parts)
- Instrumentation

Superstructure

- Design (frame, slipform core plus frame, large panel)
- Type (precast concrete, cast-in-situ reinforced concrete, timber, steel, composite)
- Geometry
- Start date and timescale of construction
- Construction technique (slipformed, bolted frame, composite)
- Dead and live loading (design code used)
- Design and regulatory codes adhered to at the time of design
- Design specification of materials, current condition

Exposure

Finding information about the exposure of the building is crucial for the reuse of concrete members. According to BS EN 206-1, environmental actions have to be considered by determining exposure classes concerning chloride attack (XD) from de-icing salts in car park areas and chemical attack (XA) from rain or groundwater. The accurate determination of these exposure classes in the planning process is the basis for a durable construction. Investigated piles for reuse should be evaluated considering the requirements of BS EN 206-1 and how far an old structure may meet the demands of durability. This includes chemical analysis of the groundwater. Further information can be found by investigation.

6.2.5 Assessment of the information and planning of next stages of work

The information collected during the desk study needs to be collated and grouped into the subject areas in the *Required information* box in Figure 6.2. A full checklist for the collation of data is given in Figure 6.3. The completed checklist will indicate the completeness of the available information and show where more information is required to mitigate risk.

6.3 Physical investigation of foundations for reuse

6.3.1 General reuse investigation context

The desk study may have established many but not all of the parameters required. It will be necessary to obtain missing parameters and to establish confidence in the validity of the desk study findings. This section sets out the main methods of obtaining further information. It deals with investigation of geometry, integrity and material properties of piles and slabs for the most common materials (concrete, steel, timber) and the subsoil, and addresses some of the issues raised in *Section 3.3*. Measurement of performance of foundation elements (load capacity and relative movement during deconstruction of the previous structure or as part of a load test of existing piles) is discussed in *Section 6.3.4*. The most important starting material in foundation performance is the ground.

Soil

The soil and its variability are the most important factors in the reuse of foundations (Box 6.2). Information may be needed for verification of capacity of elements for reuse, their durability or for design of additional new piles and modelling of ground–structure interaction. A good understanding of the soil and foundation behaviour is vital, particularly where

Information required		Information known			
		None	Generic	Outline	Detailed
Site	Local topography				
	Current use				
	Previous use				
	Use of adjacent plots				
	Instrumentation in adjacent plots				
Ground	Geology				
	Ground water				
	Archaeology				
	Underground construction				
	Services				
	Instrumentation				
Foundations	Type				
	Design				
	Design loads				
	Geometry				
	Installation date				
	Construction method				
	Materials specification				
	As built information				
	Foundation tests				
	Condition				
	Instrumentation				
Superstructure	Type				
	Design				
	Geometry				
	Construction technique				
	Loadings				
	Design codes used				
	Materials specification				
	Condition				
	Instrumentation				

Figure 6.3 Checklist for desk study (this list is not exhaustive but shows the principal information required)

Box 6.2 The importance of the soil in foundation reuse

Poulos (2004) quotes Terzaghi. In his 1951 research paper to the Building Congress, Terzaghi stated:

'Modern methods for the design of pile foundations...can be considered reliable, provided the subsoil of the proposed foundation has been adequately explored. On the other hand, if the subsoil conditions have been misjudged, for instance, on account of inadequate sampling operations, erroneous interpolation between boreholes, or lack of care in the examination of the soil samples, the difference between anticipated and real settlement can be distressingly important.'

It is the ground that dictates the type of foundation and by association the ground (and groundwater) which influences likely types of defects. The relative homogeneity (or lack of it) of ground across a site is probably one of the major issues in reuse and this needs to be established in the form of a ground model.

strain compatibility is needed across new and old elements. The investigations should aim to determine the ground properties that will be most crucial over the ranges of stress and strain that are relevant to the reuse. In practice, when mixing new and old foundations movement or strain compatibility will generally be crucial. While the techniques for investigating soil properties are well-established, the requirements may be more onerous than in a traditional design situation.

Foundations

Having established a ground model (see Box 6.2), it is essential to investigate the material integrity of foundation elements and their interaction with the ground. Defects can occur or be introduced at different stages during the foundation lifetime including:

- when installed (poor workmanship, unexpected ground conditions),

- under the original loading (environmental attacks, changes in groundwater),
- during demolition of the original superstructure and removal of ground,
- connection works between reused element and new superstructure.

If the assessment has revealed any information that one of these defects might affect the foundations, possible damage scenarios should be developed and appropriate investigations must be carried out.

Confidence levels need to be high concerning avoidance of brittle failure. This means that even in the most ideal of reuse situations (ie perfect records, good trading practices, known and reputable contractors, re-warranty certificates and no historical anomalies in building performance) some tests will still be required. The further away from the ideal situation, the more likely the requirement for more tests. The economics for obtaining this information can outweigh the financial advantages of reuse, other drivers (archaeological, sustainability) for continuing need to be evaluated (see *Chapter 2*).

6.3.2 Investigation techniques for foundation elements

Tests will comprise intrusive testing and non-destructive testing (NDT). Intrusive examination will require costly sample recovery and testing (eg concrete pile coring with cylinder crushing and timber pile pullout with chemical testing (see *Section 6.5* for case study examples). NDT, given access and indexing to physical testing to provide confidence can be used to verify integrity or investigate anomalous areas reasonably economically.

Conventional intrusive testing is well understood. This *Handbook* concentrates on NDT and its current capabilities and limitations in the reuse situation. However, NDT development is rapid. It will be necessary to keep up-to-date with the latest techniques to ensure that the optimum NDT approach is adopted (Schickert et al 2003).

6.3.3 NDT investigation methods

This section sets out the main testing challenges for investigating foundations for reuse. Appropriate NDT methods in civil engineering (NDT-CE) are described.

6.3.3.1 Requirements for a successful investigation

Investigations by NDT-CE methods should not be separated from the rest of the assessment and design process. All efforts have to be integrated into the overall procedure. The testing programme should be outlined in cooperation with client, designer and contractor. All partners should be aware of capabilities and limitations of the applied methods. There should be agreements on how to change or extend the testing programme if the need occurs and how to assess and use the results. The possible consequences of unfavourable results (eg negative test results of pile integrity testing) can (and should) be integrated into the assessment and design plans (as illustrated generically in *Chapter 3* and in practice by Vaziri 2005). The investigations specialist should be available for consultation after delivery of his report.

Knowledge of expected foundation type, location and geometry plus soil structure and properties is essential for an effective selection of investigation methods and parameters. The consultant and the designer should specify in as great a detail as possible the foundation parameters to be investigated.

Most investigation methods need at least partial access to the foundation element to be assessed. For example, the pile heads (or the shafts near the head) have to be accessible for pile length measurements by the low strain integrity testing method. Screed or sealing layers have to be removed in basements to check foundation slab geometry by, eg the ultrasonic echo technique. This may lead to additional costs or delays in the construction process, which have to be taken into account. However, such costs and delays can be minimized if the investigation is properly planned and well-integrated in the demolition/construction process.

NDT methods need calibration against prior knowledge, against material/technique variation with response and against validation parameters. For example, length measurements by low-strain pile integrity testing are influenced by concrete quality. Often concrete quality will not be known with any certainty so use of a complementary method of length determination should be considered for confirmation (eg parallel seismics). Alternatively, concrete quality could be determined via coring and the cores indexed via wave velocity measurement. Calibration requirements and costs need to be accounted for in the reuse decision process (see *Chapter 5*).

6.3.3.2 NDT for piles and slabs

Foundations have to be assessed for:

- *Geometry:* length, diameter, thickness, pile shape, location of piles beneath slabs, reinforcement size and location, concrete cover;
- *Integrity:* construction faults, cracks, material quality, corrosion, reinforcement corrosion;
- *Load capacity:* static and dynamic pile load test (see *Section 6.3.4*).

Tables 6.1 and 6.2 summarise the most common NDT techniques for assessment of foundations in the reuse process and Tables 6.3 and 6.4 their limitations and hence their most appropriate application.

For some of the parameter investigations in Tables 6.1 and 6.2 several methods may be possible. Results depend on the boundary conditions. Tables 6.3 and 6.4 give criteria for whether a method is good, less appropriate or inappropriate for use. Values in brackets are effectively inappropriate for practical use.

6.3.3.3 Description of applicable NDT methods

The following sections present advanced geophysical methods that have been intensively developed over the last decade. Only the most commonly used methods are described.

Table 6.1 Appropriate testing methods for slab foundations in the reuse situation

Required parameter	Most appropriate technique	Access requirements, applications, limitations and remarks	Alternative methods
Slab thickness	US echo To improve results: reconstruction calculation	US data has to be measured direct at the slab surface. No extra layer above the concrete or air-filled layer (eg sealing) within the slab. Several measurement points to avoid misinterpretation. Rough surface, several concrete layers, high reinforcement ratio impede the measurement. Attainable thickness: not reinforced: > 2 m; reinforcement (2 layers ∅28, 10 cm spacing, crosswise): 75 cm.	Impact-echo: up to 1 m Radar: 30–60 cm
Different layers	US echo To improve results: reconstruction calculation	Every layer causes reflection that may hide information from greater depths. In case of even a very thin air-filled layer (eg sealing) a total reflection does not allow any information beyond that layer.	Radar
Location of piles and strip foundations (beneath slab)	US echo with reconstruction calculation of measured data	US data has to be measured direct at the slab surface above the object. No extra layer above the concrete. Measuring grid: 2–10 cm necessary; no single point measurement. A joint between object and slab, high reinforcement ratio and slab thickness > 70 cm may impede the location of the object.	Excavation
Location of voids	US echo with reconstruction calculation of measured data	US data has to be measured direct at the slab surface above the object. No extra layer above the concrete. Measuring grid: 2–5 cm necessary; no single point measurement. Honeycombs (10 × 10 × 10 cm^3) could be detected between 10–25 cm. Decreasing reliability with increasing reinforcement ratio.	Impact-echo Radar Coring
Concrete quality	Rebound hammer	Compressive strength of surface near concrete by evaluating rebound results: concrete surface has to be accessible.	Coring, compressive strength
State of corrosion	Potential mapping	Concrete surface has to be accessible. Reinforcement excavated at one spot. Only areas of active corrosion can be detected. Commercially available devices allow investigation of several hundred m^2 a day. To measure loss of section of the rebars, excavation is necessary.	Excavation, coring
Reinforcement ratio Concrete cover	For near-surface reinforcement: devices working with eddy current and imaging (cover meter).	Concrete surface has to be accessible. Commercially available devices offer functions to measure the concrete cover and image rebars. For more than one reinforcement layer US and radar should be used. US with reconstruction calculation allows imaging of the lower reinforcement layer in a slab up to 30 cm.	For deeper reinforcement: US echo or radar and reconstruction calculation

US = Ultrasonic

Further testing methods are described at www.bam.de/zfpbau-kompendium.htm

- *Ultrasonic-echo* for thickness measurement and location of objects,
- *Radar* for the detection of metal reflectors (reinforcement),
- *Low-strain testing* for pile length measurement and pile integrity testing,
- *Parallel-seismic* for pile length measurement as a reference method for low-strain testing,
- *Mise-à-la-masse* to locate the reinforced part of a pile and the length of steel piles.

Guidelines for carrying out various NDT tests are given in *Appendix D*.

6.3.3.4 Concrete test hammers and cover meters

Concrete test hammers (Figure 6.4 a) have developed over several decades and represent a well-established NDT practical tool. They obtain the near-surface concrete compressive strength by measuring the hardness (rebound value, R) of the concrete surface and converting this into compressive strength using conversion curves. Various versions of testing hammers either with paper chart recorders or digital display with automated data recording and

Table 6.2 Appropriate testing methods for pile foundations in the reuse situation

Required parameter	Most appropriate technique	Access requirements, applications, limitations and remarks	Alternative methods
Pile length Load capacity (indirect): sufficient pile length	Concrete and timber: low-strain method Steel section: parallel seismic or mise-à-la-masse (see below)	For piles with *access to the pile head or pile shaft* by excavation. Pile length L: 5–20 m; L/∅ < 30. Standard testing device for concrete available; possible for timber piles but less reliable. Information about subsoil increases reliability. Length measurement for piles under slab not possible. To assess load capacity: comparison with recordings from subsoil if load-bearing layer is reached.	Parallel seismic
Diameter (below slab)	Only for concrete piles: US echo with reconstruction calculation of measured data	US data has to be measured at the slab surface above the pile (ie no extra layer above the concrete). Measuring grid: 2–10 cm necessary; no single point measurement. A joint between pile head and slab, high reinforcement ratio and slab thickness > 70 cm may impede the determination of the diameter.	Excavation
Pile shape/ Load capacity (indirect) Construction faults Cracks	Only for concrete piles: low-strain method for pile integrity testing	For piles with *access to the pile head or pile shaft* by excavation. Pile length L: 5–20 m; L/∅ < 30. Standard testing device for concrete available. To assess load capacity and reveal construction faults: deviation from expected pile shape (diameter changes), certain shape of pile toe (eg Franki pile), cracks (not parallel to pile axis) detectable. Non-analytical method (different interpretations can arise from a significant signal. Therefore detailed information about subsoil and expected pile shape necessary.	Coring, excavation cross-hole sonic logging (CSL): holes for US-testing in the existing pile needed. (Not given for old constructions.)
Material quality	Concrete and timber: low-strain method	For piles with *access to the pile head or pile shaft* by excavation. Pile length L: 5–20 m; L/∅ < 30. Significant deviation of expected wave propagation speed indicates poor concrete or rotten timber. Non-analytical method (see above)	Coring, excavation Pullout and testing (particularly timber)
Pile reinforcement	Only for concrete piles: mise-à-la-masse or induction method	Access for boreholes 0.5 to 2 m beside the pile needed. The borehole must be lined with a plastic and slotted casing to ensure electrical contact between the measuring cable and the ground. At the top of the pile, access to the reinforcement is required. Electrical continuity of the reinforcement is required to determine lengths of reinforcement where multiple cages are used.	
Corrosion	No reliable test method known.	Potential mapping is only suitable for measurement at concrete surface.	Coring, excavation

US = Ultrasonic

Further testing methods are described at www.bam.de/zfpbau-kompendium.htm

compressive strength calculation are available (10–70 N/mm^2).

Similarly, light, practical and often used are cover meters to identify near-surface steel occurrence and position (Figure 6.4). Modern cover meters use eddy currents combined with pulse induction techniques and give very reliable results. Advanced devices allow imaging of the results of a tested area.

6.3.3.5 Ultrasonic-echo

A transmitter generates ultrasonic impulses (25–200 kHz) as shown in Figure 6.5 (a). The reflection off the backwall or a defect is received on the same side of the concrete member. Areas of bad concrete quality may lead to an increased transit time ('backwall shift') and weakened signal (Figure 6.5 b). In case of a defect with different acoustic properties (E modulus, density) compared with the sound concrete, a defect might be detected directly from its reflection or

Table 6.3 Limitations of several NDT methods (slab foundations)

Construction feature (for slabs)	NDT method US-echo	Impact-echo	Radar	Potential mapping	Core drilling
Reinforcement ratio					
Unreinforced	+	+	+	–	+
Low	+	+	o	+	o
High	o	+/o	o/–	+	o/–
Water content					
Low	+	+	+	+	+
High	o	o	–	o	+
Multi-layer systems					
Screed/concrete	o	o	+/o	o/–	+
Thermal insulation/concrete	–	–	o	(–)	o/–
Sealing layer included (dry)	–	–		(–)	–
Sealing layer included (wet)	–	–	–	(–)	–
Layer with good conductivity (eg steel plate)	+/o	+/o	–	(–)	–

US = Ultrasonic

+ = good, o = less appropriate, – = inappropriate, ( ) = effectively inappropriate for practical use

Table 6.4 Limitations of several NDT methods (pile foundations)

Feature	NDT method Pile integrity testing	Parallel-seismic	Mise-à-la-masse	Core drilling
Accessibility				
No access to pile head	o[1]	o[1]	+	o
No access to shaft	o[1]	o[1]	o	+
No access to reinforcement	+	+	–	+
Slab/cap on piles	o	o	o	o
Construction				
Metallic	o	+	+	–
Reinforced	+	+	o	+
Not metal parts	+	+	–	+

+ = good, o = less appropriate, – = inappropriate

[1] Access either to head or shaft near head required

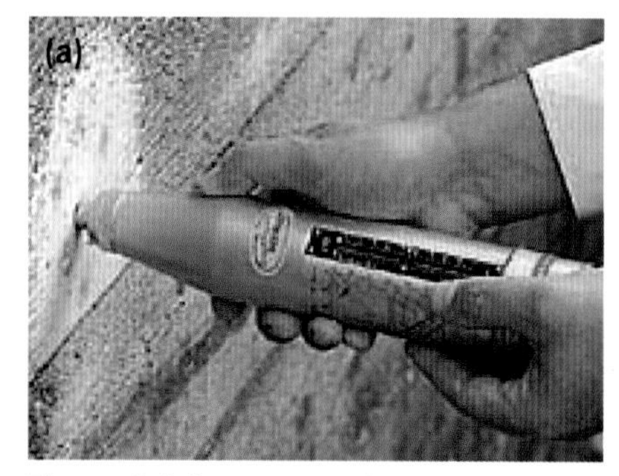

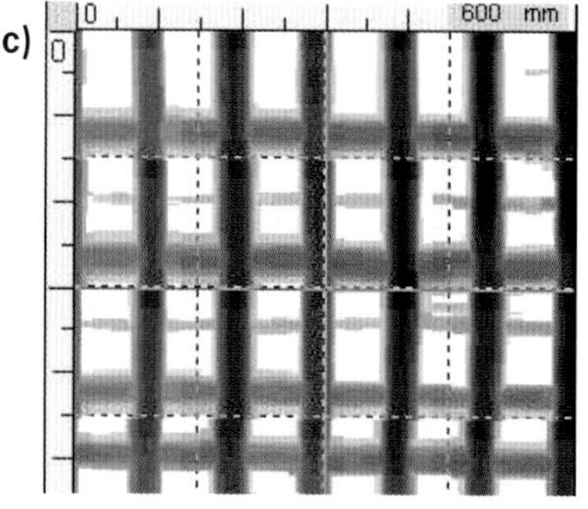

Figure 6.4 Concrete test hammers: **(a)** concrete test hammer with direct display; **(b)** rebar location by scanning a surface with a sensor; **(c)** imaging of the located rebars in an area of 60 × 60 cm^2

indirectly by shading the backwall ('shadow effect') as shown in Figure 6.5 (c).

For all cases shown in Figure 6.5 (a)–(c), measurements along a line or a grid have to be taken. Figure 6.5 (d) illustrates instrument types to accommodate structural configuration best. Both systems operate with sensors that do not require coupling agents (dry-point contact sensors). Provided measurements are taken along a line or grid using advanced data processing (Schickert et al 2003; reconstruction calculation), reflections can be imaged in longitudinal sections or sections parallel to the surface.

Air-filled (eg sealings) or debonded (eg screed) layers cause total reflection; no information can be gained beyond that point. Reinforcement, poor concrete quality and rough surfaces can decrease the quality of results.

Low frequencies (25 kHz) allow greater penetration depth for thickness measurement and imaging of geometry. Figure 6.5 (e) shows a longitudinal section with the reflections

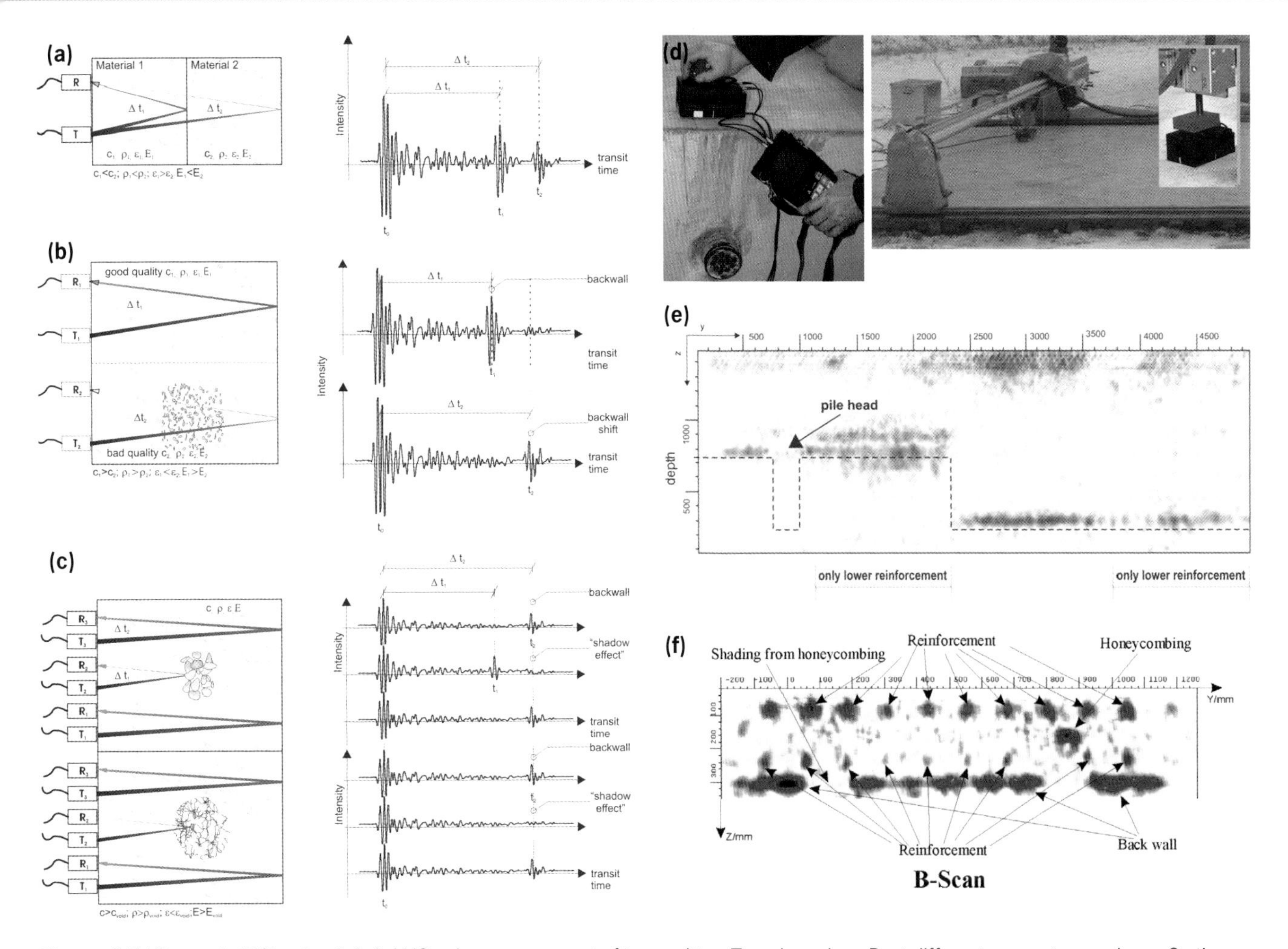

Figure 6.5 Ultrasonic (US)-echo: **(a)–(c)** US-echo arrangement of transmitter, T, and receiver, R, at different concrete members. On the right side, a typical but idealised recording is shown. **(d)** Measuring devices: hand-held and automated data recording. **(e)–(f)** Imaged results after reconstruction calculation: **(e)** foundation geometry; **(f)** rebars and voids

imaged over the depth of 75 cm and 125 cm thick reinforced foundation slab. The backwall of both depths can be detected as well as the location of pile heads beneath the slab (interrupted backwall reflection) (Taffe et al 2005).

High frequencies (85 kHz) allow low penetration but better resolution of single rebars and defects (eg honeycombing) in near-surface areas. In Figure 6.4 (f) the lower reinforcement layer of a 30 cm thick reinforced slab become visible (Krause et al 2003).

For further details and guidance for a successful application of this technique refer to *Appendix D* or DGfZP (1999). The whole testing process of ultrasonic-echo regarding construction, history and possible deterioration is summarised in the 'Testing protocol' for foundation slabs in *Appendix D*.

6.3.3.6 Radar

Radar involves the generation of electromagnetic impulses and the recording of their reflections and backscattering at layer boundaries and objects (Figure 6.6 a). Sections with a higher moisture content lead to longer transit times, which become apparent by a so-called 'backwall shift' (Figure 6.6 b). Total reflections occur at metal objects (Figure 6.6 c). Figure 6.6 (b) shows how the typical hyperbola shape from a metal reflector emerges by measuring from several positions. In radargrams the position of a rebar is marked by the top of the hyperbola.

Normally, a hand-held antenna-containing transmitter and receiver is moved along the surface (Figure 6.7 a). Location is recorded by a positioning wheel and the radar traces (time series of signal at receiving antenna) can be presented in longitudinal sections, ie radargrams (Figure 6.7 b) or sections parallel to the surface, ie time slices (Figure 6.7 c).

High frequencies (1.5 GHz) lead to high resolution (cm-range) but low penetration depth (up to 40 cm for thickness measurement in reinforced structures). Low frequencies (500 MHz, 900 MHz) have a higher penetration depth up to 100 cm but this rapidly decreases with increasing reinforcement ratio. Moisture also adversely affects the result quality. Hand-held devices allow several hundred square metres to be investigated in a day. Radar can be used when the penetration depth of cover meters is exceeded or more than one layer of reinforcement needs to be detected.

For further details and guidance for a successful application of this technique, see *Appendix D*. The testing process of radar regarding construction, history and possible deterioration is summarised in the testing protocol for foundation slabs in *Appendix D*.

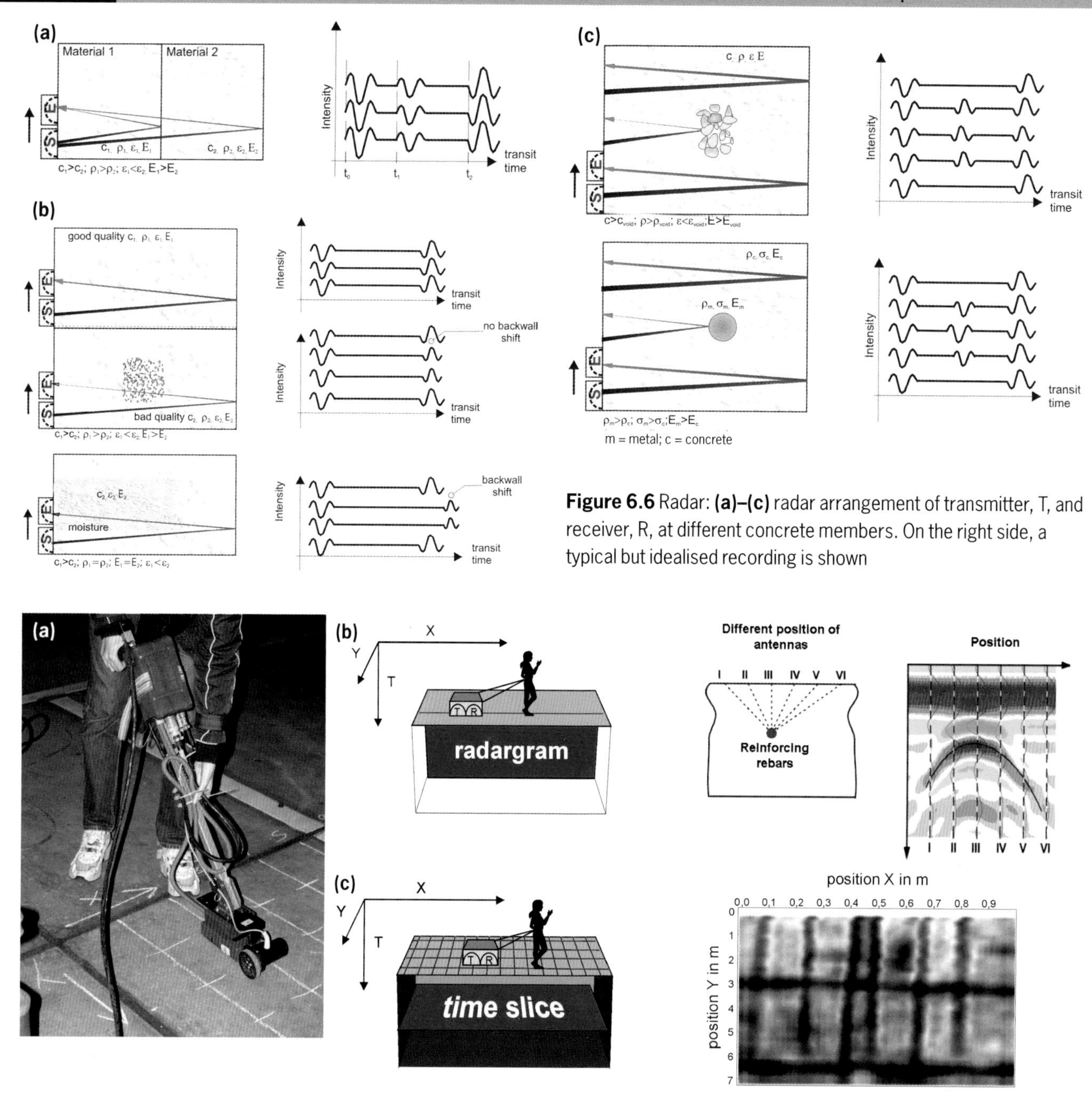

Figure 6.6 Radar: **(a)–(c)** radar arrangement of transmitter, T, and receiver, R, at different concrete members. On the right side, a typical but idealised recording is shown

Figure 6.7 Radar measurement: **(a)** radar measurement with hand-held antenna and positioning wheel; **(b)** longitudinal section of recorded data along a line (radargram). The typical hyperbola shape of a single rebar becomes obvious (DGfZP 2001); **(c)** section parallel to the surface showing reflections from rebars in a certain depth (time slice) recorded along a grid (Krause et al 2003)

6.3.3.7 Low-strain method

The low-strain method is often referred to as pile integrity testing (PIT). A velocity sensor is placed on top of the pile and a force impulse is generated by a hammer blow as shown in Figures 6.8 (a) and (b). The sensor records the resulting movement of the pile head (Figures 6.8 c–f). If the pile head is accessible and the concrete surface is prepared, a measurement is done in a few minutes.

The length of the pile or the depth of a major defect can be calculated by multiplying the time difference t_f between impact and reflection by the velocity of sound in concrete (3500–4200 m/s in most cases). Reflections are caused at major impedance changes which are present at major increases or decreases of cross-section and/or concrete quality, and at the pile toe if the acoustic parameters of the surrounding soil differ significantly from the pile concrete. The shape of the reflection together with information about the subsoil gives hints for the interpretation as shown in Figure 6.8:

(c): sound pile with toe reflection of depth of 7 m,

(d): void of depth of 5.5–6.0 m. The toe reflection is not visible. If the planned pile length is not known, this defect could be misinterpreted as a pile too short.

(e): in case of loss of cross-section of depth of 5.5–6.0 m the pile toe reflection keeps visible.

(f): in case of a defect in the upper 2 m the reflection could easily have been interfered with by the impact signal.

Figure 6.9 (a) shows three different piles with an impedance change. The impedance change occurs either from:

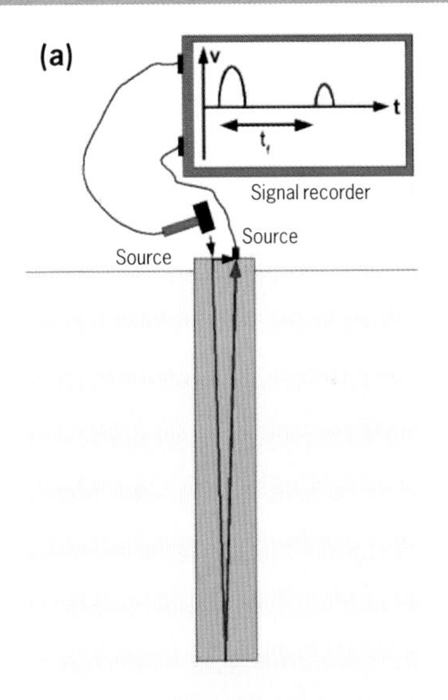

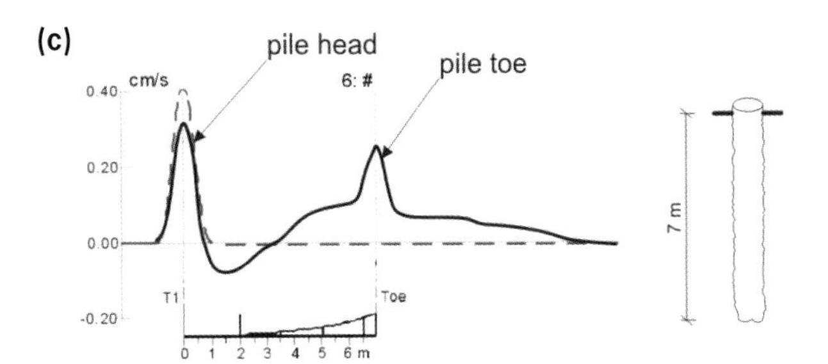

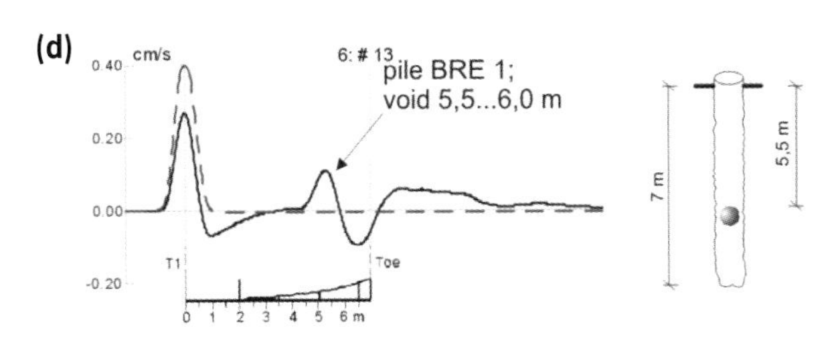

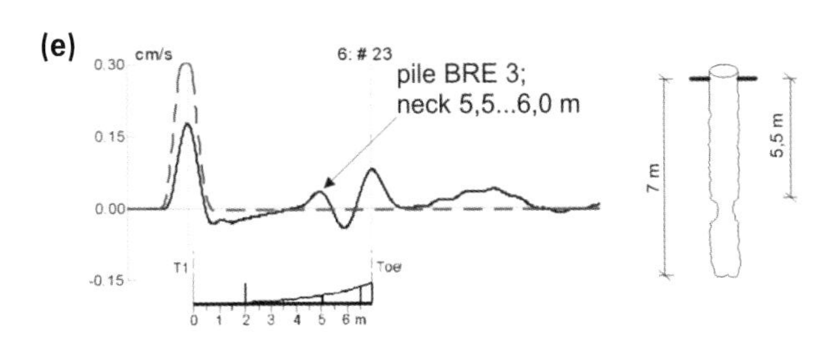

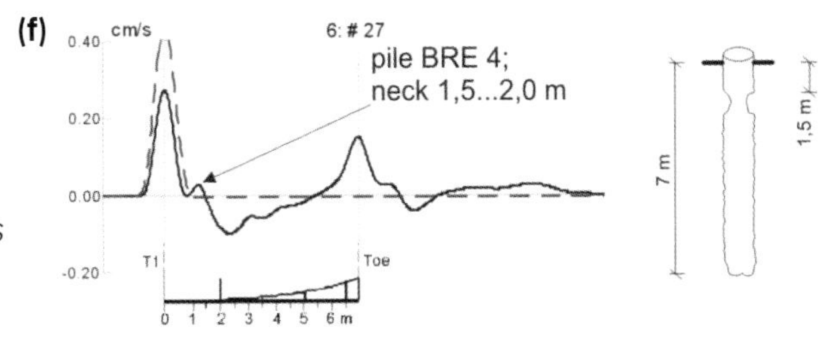

Figure 6.8 Low-strain method: **(a)** principle of low-strain method; **(b)** pile testing device with sensor and hammer at the pile top; **(c)–(d)** typical recordings of reflections measured at the pile top *vs* depth (transit time)

- loss of section,
- reduced density or E modulus of the concrete,
- soil resistance change.

However, the velocity time record is in all three cases quite similar. This illustrates the importance of the boundary conditions and the need for expert and experienced interpretation in gauging reasons for impedance change.

If the pile length is totally unknown (which is often the case in a reuse situation), a reference test method such as parallel seismic should be applied to avoid misinterpretation. An example of how this could occur is well-illustrated in the Chattenden trace interpretation (Figures 6.9 b and c). Trace b could be interpreted as a 10 m pile with a void at 5 m depth when in fact it is a sound pile 5 m long. The results from Chattenden are further discussed in Box 6.3.

In the case of homogeneous ground, pile length can be indexed to pile capacity and so the use of NDT can minimise the related costs of expensive load tests to verify capacity.

The testing process of low-strain method regarding construction, history and possible deterioration is summarised in the testing protocol for foundation slabs in *Appendix D*.

For pile integrity testing (German Society for Geotechniques 1998) gives a system for classification of tested piles. Four possible classes of results should be differentiated:

- *Class 1:* the pile is intact,
- *Class 2:* the pile is not intact and has a minor quality reduction,
- *Class 3:* the pile is not intact and has a major quality reduction,
- *Class 4:* the signal is unclear (eg due to cracks, concrete contamination at the pile head).

6.3.3.8 Parallel-seismic

Parallel-seismic is an alternative acoustic method for measuring pile length. It can be used as a reference method for the low-strain method when the pile length of a structure is totally unknown. A hammer blow on top of the pile generates elastic waves. Transit times through pile and soil are measured using a sensor (geophones or hydrophones) in a borehole at different depths (see Figure 6.10). First arrivals are picked from the records and are displayed in a graph (time *vs* depth). At the pile toe, the slope of the resulting curve changes. The length of the pile is determined from this point of inflection or (more accurately) by model calculations.

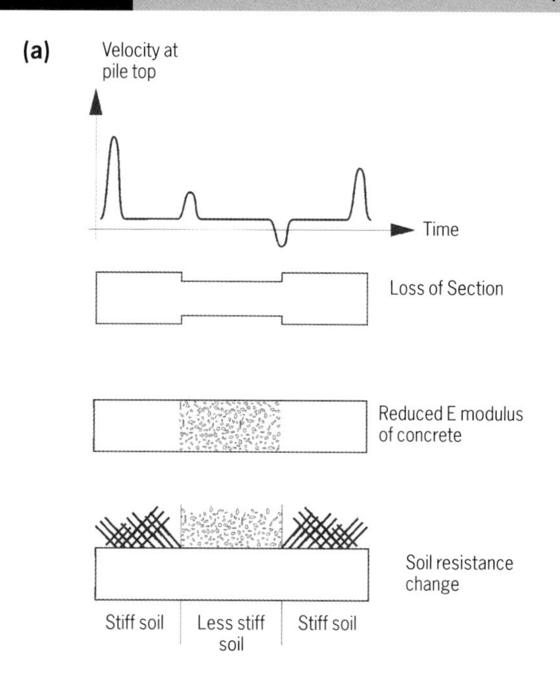

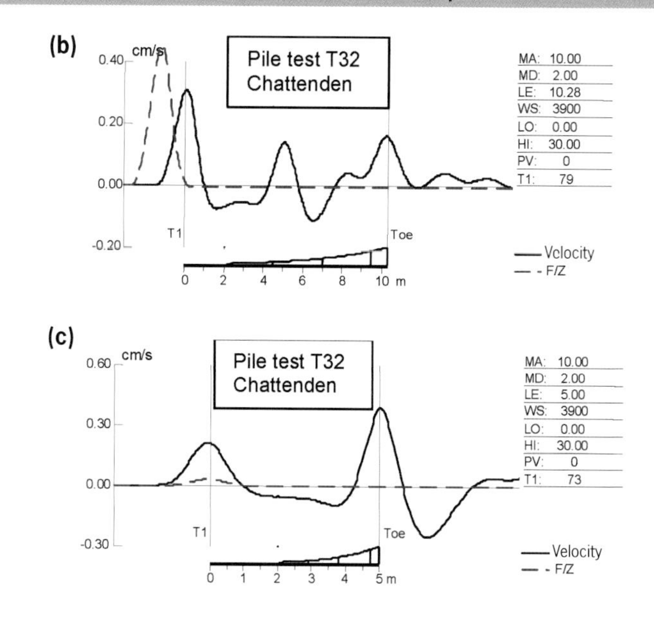

Figure 6.9 (a) Idealized time/velocity recording representing three different pile situations; **(b)** velocity time record of a pile 'expected' to be 10 m long. The expected toe echo is the second (multiple) reflection of the true pile toe at 5 m. The estimated 'void' at 5 m depth is the (true) pile toe; **(c)** corrected velocity time record of the same pile

Box 6.3 Pile length measurement at Chattenden

Pile length measurements were carried out by a RuFUS partner, BAM, at the test site in Chattenden, UK, where the ground conditions were uniform and well-understood. They illustrate the effects of the pile geometry, composition and installation technique on the pile length measured using the low-strain method. Initial measurements of pile concrete wave velocity (3,900 m/s) were carried out on a 10 m long straight shafted bored reference pile. A blind prediction of pile length was then made for:

- straight shafted bored piles constructed by 3 different methods **(a)**,
- screw (helical) displacement piles **(b)**,
- straight shafted piles with a steel casing over the upper portion of the pile **(c)**.

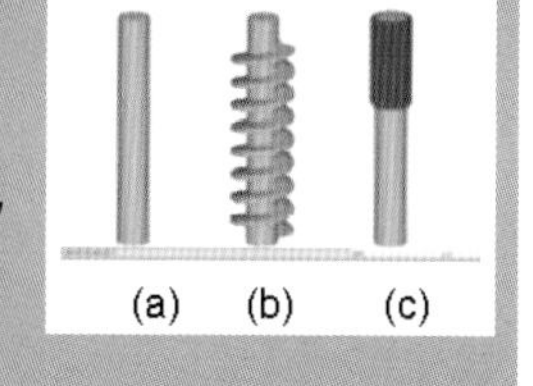

The results from these blind predictions showed the following:

(a) Similar wave velocities occurred in the straight shafted piles, irrespective of the construction method and so length predictions tallied well with constructed lengths.

(b) Screw (helical) displacement pile lengths were over-predicted (by 8–12%) because the wave velocities were slower than for the reference piles. Wave speeds for screw displacement piles need to be indexed against a reference screw displacement pile of known length in the same ground conditions.

(c) Predicted pile lengths for piles with steel casing over the upper portion of the pile were shorter by (6–15%) than the constructed lengths because the wave velocity was higher than for the reference pile. The steel casing reduces the transit time so composite steel/concrete wave velocity should be adopted for piles of this type. In addition, the user must be careful not to interpret the observed phase shift of the wave at the base of the casing as either an increase in pile diameter or a change in subsoil.

The borehole has to be cased (with bottom end cap and grouting of the space between casing and soil) and, if hydrophones are used, filled with water. The distance to the pile should be less than 3 m and the borehole length should exceed the pile length by three times the distance to the pile. An alternative approach is to use a seismic cone penetrometer (SCPT) (Butcher et al 2005, Butcher 2006, Rankka & Holm 2006) as the receiver of the signals. The SCPT can be jacked into the ground to the required depth and the hammer blow made on the pile head. The SCPT can be advanced to a new depth and the procedure repeated. This technique avoids the need for the lined borehole but relies on the ability of the equipment to jack the SCPT to the required depths.

For this method only, France has a national standard (NF P 94-160-3; AFNOR 1993). Recommendations are available in the UK from CIRIA (Turner 1997) and the USA (Federal Highway Agency 2003).

6.3.3.9 Mise-à-la-masse

Mise-à-la-masse is a method of determining the length of electrical conductive elements in the soil, such as steel piles or reinforcement in concrete piles. Mise-à-la-masse is therefore likely to be applicable to all piles that are metallic or contain metallic elements. The length of reinforcement can be determined if there are multiple cages down the length of the pile and electrical continuity between the cages.

Figure 6.11 (a) illustrates the mise-à-la-masse principles. A conductive element of the pile must be available in order that an electric potential can be induced via the reinforcement. A remote earth point has to be used (B in Figure 6.11 a), a typical distance of 50 m is enough for 10 m long piles. A series of electrodes (M) is inserted in the borehole, parallel to the pile. These are used to log the voltage variation with depth.

Figure 6.10 Parallel seismic: **(a)** principle of parallel-seismic (show 3 × distance to pile to borehole depth); **(b)** method application on site: hammer blow on pile top and sensors in a borehole; **(c)** depth of the sensor *vs* transit time measured in the sensor position. The inflection point marks the position of the pile toe

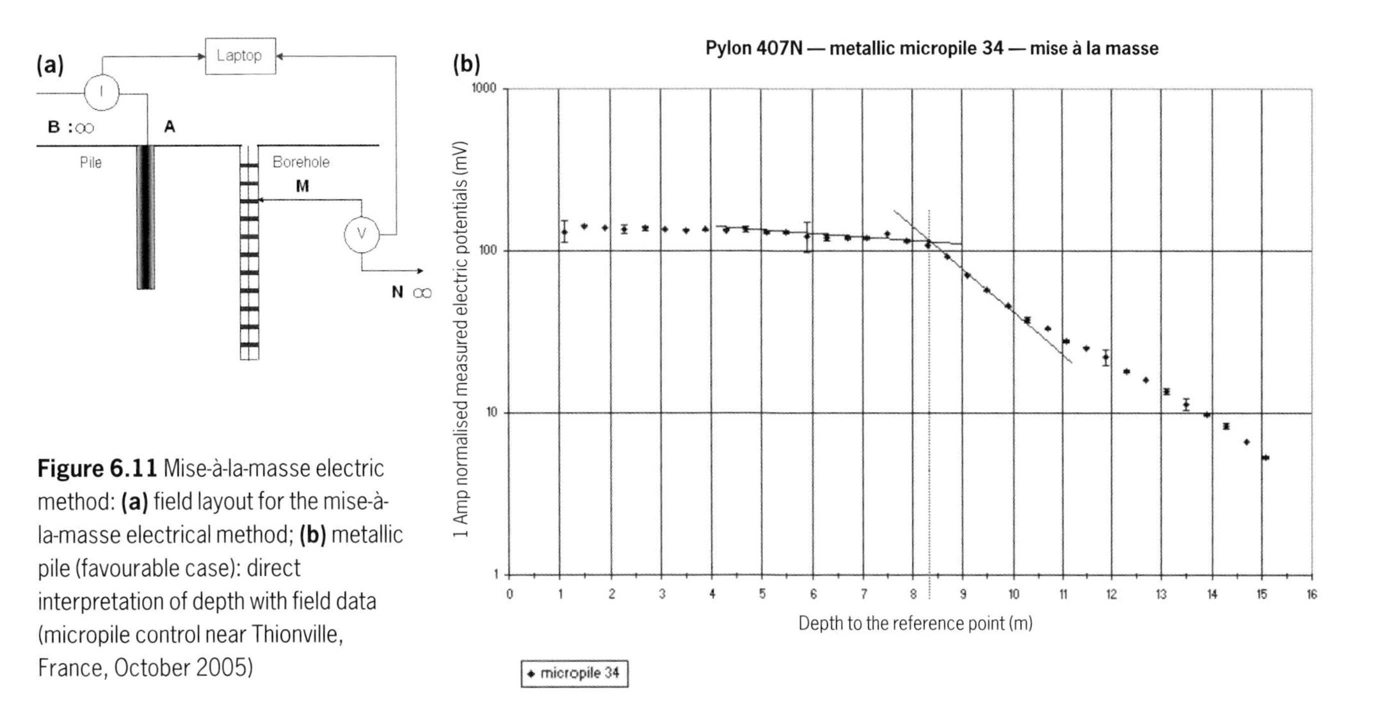

Figure 6.11 Mise-à-la-masse electric method: **(a)** field layout for the mise-à-la-masse electrical method; **(b)** metallic pile (favourable case): direct interpretation of depth with field data (micropile control near Thionville, France, October 2005)

With metallic piles, an approximate, direct interpretation of the pile length can be performed on the field measurements: the end of the pile is associated with a break in the potential slope when represented by a logarithmic scale. Accuracy on a recent case study (see Figure 6.11 b) has been assessed to be approximately 5% (this accuracy can be enhanced by further field calibration).

For reinforced concrete piles, reinforcement length can be determined by comparing the obtained field potential curve with a data bank of models obtained from classic geophysical surveys. Pile length need not correspond to pile reinforcement length.

6.3.4 Performance

6.3.4.1 Deconstruction

The deconstruction of an existing building is an ideal opportunity to collect information about the structure, its foundation and the ground that has supported it during its life. The loads in a structure will be redistributed during the lifetime of a building and may be different from the loads calculated in the design. The design loads of the structure will have included factors to account for dead and live loading that may or may not have been fully utilised during the life of the building.

Why monitor deconstruction?
The behaviour of a building foundation during deconstruction can be monitored and will be similar to the unloading cycle of a foundation load test. Measuring and understanding the behaviour during unloading will lead to a better understanding of both the actual loads mobilised during the life of the building and the likely reloading behaviour.

Monitoring can give:

- a good idea of the working loads on the foundation for comparison with the design loadings,
- a picture of the behaviour of the foundations and surrounding ground for comparison with the original design philosophy.

Robust instrumentation installed to sufficient depth can:

- check on unloading behaviour of deep piles,
- be used to assess any likely tension loads held in the piles after unloading, and
- determine where the significant movements have occurred.

Butcher et al (2006b) shows vertical movements measured on a deconstruction site. The instrumentation was installed to measure vertical movements of the ground close to the piles and, more significantly, at depths beneath the bottom of the piles. There was significant movement beneath the pile base and the movements extended to a depth below the pile bases equivalent to nearly twice the width of the foundation.

With care over the location of the instrumentation, it can be used during construction of the new structure to assess foundation loading and in-service behaviour (see *Chapter 7*).

Monitoring methods
Instrumentation needs to be simple and robust to survive demolition activities but have sufficient resolution as the movements to be measured will be small. Instrumentation should also be easy to repair and recalibrate should it be damaged during the demolition/deconstruction. The sensors chosen will depend on the property to be measured (see *Chapter 7*). If the loads transferred from the superstructure to the foundations are required, strain gauges may be fixed close to the base of the columns. Vertical movements across the basement floor slab between piles may be required to assess differential heave during unloading for construction of a deepened basement.

Deconstruction/demolition is a hazardous activity and preservation of instrumentation is not the first consideration of operatives. In general, instrumentation installed to monitor foundation performance will survive well until the deconstruction reaches the basement level. In many cases, it will be advantageous to use monitoring techniques that can be continued through construction of the new structure and show the effects of reloading as the new building is constructed and in-service behaviour.

The configuration of the instrumentation will depend on the type of foundation to be monitored, the distribution of stress in the ground resulting from the foundation loads and the likely movements. For example, if the existing foundation is a raft supported by friction piles the movements of the raft and the movements in the ground beneath the bottom of the piles, to a depth approximating to the width of the raft, will need to be monitored. An example of such a scheme can be found in the Stonebridge Park case study (Butcher et al 2006b).

Interpretation of data
The monitoring data will need to be interpreted in conjunction with the original foundation design to assess whether the unloading behaviour was as expected. It is likely that if the foundation was working well within capacity, the movements will be small and elastic. However, the distribution of movements could highlight different behaviour to that expected in the original design. Simplified models can simulate some aspects of pile response such as pile-bearing capacity. Yet, the most general method that can solve many aspects of pile response, including the pile settlement of a raft foundation, is high quality numerical analysis. Such analyses when applied to simulate pile response have some special features that must be considered (Potts & Zdravkovic 2001).

As the database on deconstruction expands, it is expected that deconstruction will become a more useful tool for reuse purposes.

6.3.4.2 Foundation testing (capacity) and applicability
In a foundation reuse situation, the designer will to need to have information about both the capacity and behaviour (ie load displacement) of any foundation intended to be reused. It may be necessary to know both the working load that the foundation elements can take and also the likely movements that will be generated so as to ensure strain compatibility, especially between new and old foundations.

In the following sections, methods are considered for 'testing' existing piles to assess their performance in 'reuse'. If the piles to be tested are intended for reuse it may be unacceptable to 'fail' the piles. However, if an assessment of ultimate capacity is required it may be necessary to select sacrificial piles that can be failed with performance checks being undertaken on other piles

There are various forms of load testing available to the designer to assess the likely performance of both old and new piles. The forms of testing include static tests, dynamic tests and kinematic or rapid load tests as well as variations on these.

Method selection will depend on what is required, ie performance to a 'working load' or assessment of 'ultimate capacity' or 'failure load'.

In reuse situations, all methods currently available will require access to the isolated pile head.

Static load test
The static load test (Figure 6.12) has been the most commonly used method for determining the load capacity of piles for many years and all piling engineers are generally familiar with it. Usually, it is used to assess the load-

Figure 6.12 A typical static load test arrangement

Box 6.4 Static load test

Advantages
- Load and deformation information for the foundation element can be obtained at rates close to the construction timescale.
- If piles can be failed, a true assessment of capacity can be obtained.
- The method is well-standardised, both nationally and internationally.

Limitations
- Expensive, time-consuming, large working space is needed.
- Pile head has to be accessible.
- Reaction piles, anchors, kentledge or similar is needed (testing existing piles reacting off the superstructure may be possible).

displacement and ultimate capacity of piles at the construction stage, but obviously has applications for testing for reuse.

Static load testing requires that the pile is loaded or unloaded while measuring the load and resultant displacement. The test requires some form of reaction system to load against, either piles or kentledge for compression testing or a ground-bearing frame for tension testing; in the case of piles under an existing structure, it may be possible to use the structure to supply some form of reaction.

The load is generally applied:

- either at a rate that gives a constant rate of displacement of the pile head [constant rate of penetration (CRP) test], or
- in increments that are maintained for a fixed period of time or until the rate of displacement has slowed to a stipulated value [maintained load (ML) test].

These apply to both loading and unloading.

Suitable load and displacement measuring systems also need to be available with the displacement systems arranged so as not to be influenced by ground movements induced either by the pile being tested or the reaction system.

For sacrificial or test piles, the piles can be tested to failure determined by the load at a specified displacement or the maximum load attained if a brittle performance is observed. However, for working piles, or in the case of piles for reuse, the loading sequences must be kept within acceptable ranges which typically might be to a working load or a multiple of it. This latter type of test might be considered as an NDT.

The advantage of this type of testing is that it gives true pile load displacement performance information be it from the monitoring of existing piles during unloading or from subsequent load testing of an unloaded pile intended for reuse. However, without testing to failure or to loads higher than working load with some form of extrapolation, no information is available on the current capacity and load displacement behaviour of the piles above their tested loads. In these circumstances, normal working load design practice of using a percentage of the ultimate load cannot be used.

There are currently no techniques that can successfully predict failure loads from either loading or unloading data in the 'working' range of behaviour. The only way to make such a prediction would be to test redundant piles to failure.

Box 6.4 lists the advantages and limitations of static load testing.

Dynamic load test

Dynamic load testing has been available for a number of years and has evolved from the interpretation of measurements taken during pile driving (Holeyman 1992, Rausche et al 1995, Turner 1997). It involves striking a pile with a hammer (for driven piles it is usually the same one that is used to install the pile) and observing the resulting forces and motions recorded near the pile head. By varying the weight and drop height, different forces can be applied to the pile. For conventional hammer impact tests the duration of motion is of the order of 40 ms or less. Figure 6.13 shows a typical test arrangement.

Measurements are made by attaching reusable, high-quality sensors to the pile wall, monitoring the response and calculating the force and velocity of a stress wave imparted into the pile head by the falling weight, usually by a piling hammer. Data are collected and stored automatically on a blow-by-blow basis, either during pile installation or as extra impact loads. The accelerometers and strain transducers attached to the pile give data to enable the calculation of pile top force and velocity. The signals from the instrumentation are conditioned and processed, and the data can be further evaluated using proprietary wave equation analysis software packages

Much of the experience gathered with this method has been obtained from measurements made at the end of driving piles and then related to their capacity measured shortly after driving, as measured using static testing. These will obviously be affected by the inherent problems of the short term nature of the capacity assessment. Published data suggests that capacity assessment is possible in some soils but displacement behaviour is significantly over predicted. More research is required to better understand the test. It may be possible to utilise the test better in a reuse situation if it can be 'calibrated' for site specific application against a static test.

Box 6.5 lists the advantages and limitations of dynamic load testing.

Figure 6.13 A typical dynamic load test arrangement

Box 6.5 The dynamic load test

Advantages
- Quick, numerous tests can be achieved each day.
- It uses readily available equipment.

Limitations
- It requires analysis programs to interpret data.
- Pile head has to be accessible.
- Piling blanket will be required if large rigs are used.
- The test is much more rapid than a static test and so 'rate' effects have a significant influence on capacity.
- The higher loads required in this test compared with the static capacity require that the pile is capable of taking the loads without structural failure.
- Piles need to be failed to assess capacity.

Rapid load test

The rapid load test in its most common form loads the pile using the force produced by a rapid-burning propellant fuel within a combustion chamber (Middendorp 2000). This accelerates a mass upwards, reacting against and pushing the pile downwards. The force applied can be varied usually by varying the amount of fuel to be burnt but also when necessary by changing the weight of the mass. The test uses purpose-built equipment that must be transported to site and erected but once ready can be easily used to test several piles per day.

During the loading cycle, readings are taken of the load and acceleration or displacement and the data is stored on a PC. The mass is caught in a controlled manner by hydraulic rams, or other system, and can be lowered down within minutes to carry out another loading on the same pile if required. A typical arrangement is shown in Figure 6.14. The duration of loading is several times longer than the dynamic tests and of the order of 100–200 ms.

The rapid loading method avoids problems normally encountered with wave equation analysis by loading the whole of the length of pile simultaneously, rather than through a wave front as in the dynamic analysis. The test generates a load *vs* displacement response for the pile (the displacement being obtained via integration of the accelerations or better by direct displacement measurement).

As with the dynamic test, 'rate' effects can play a significant role in the test's interpretation, especially in clay soils (Powell & Brown 2006, Middendorp 2000). However, developments are taking place that may improve this situation. These rate effects can result in failure loads up to

Figure 6.14 A typical rapid load test arrangement: **(a)** test in progress, **(b)** computor monitoring and recording

Box 6.6 The rapid load test

Advantages
- Quick, numerous tests can be achieved each day.
- It can yield load displacement curves directly but subject to possible rate effects.

Limitations
- It requires purpose-built equipment.
- It may require new analysis methods to interpret data for rate effects in clays and silts.
- Pile head has to be accessible.
- The test is much more rapid than a static test so 'rate' effects have a significant influence on capacity.
- The higher loads required in this test compared with the static capacity requires that the pile is capable of taking the loads without structural failure.
- Piles need to be failed to assess capacity.

twice the static load being required to fully mobilise the pile; reducing these in a logical way to static capacity requires further study. At present, capacity prediction in sands and gravels seems reasonable but better understanding in silts and clays is still required. As with the dynamic test, local calibration to static test results can allow the test to be used widely across a site to verify capacity or performance. However, in clay soils current thinking is that rapid loads of twice the static working load may be required to verify the static working load because of rate effects.

Box 6.6 lists the advantages and limitations of rapid load testing.

As part of the RuFUS project, a comparison of test results from the three methods above for pile tests in glacial till showed that in terms of capacity:
- CRP results are a little higher than the ML,
- the rapid load test measured capacity is nearly twice the static capacity,
- the corrected rapid load test capacity using general current practice for correction is still some 30% above static; application of additional rate correction improved prediction,
- the dynamic tests were difficult to interpret consistently and gave results higher than static.

In terms of displacement:
- the CRP and ML results were similar up to working load,
- the measured rapid load test displacement was similar to the static behaviour,
- the 'corrected' rapid load test displacement predicted too soft a behaviour,
- the dynamic tests gave both higher and lower derived stiffnesses compared with static testing.

General guidance on all 3 methods can be found in ICE (2007).

In summary, static pile testing methods are expensive and time-consuming, but have the advantage of simple analysis and real load displacement information. Conversely, dynamic and rapid load testing methods are quick to carry out but require more specialised equipment and analysis and have yet to prove their reliability in all soils. All require full mobilisation of the pile to give a capacity.

Current published relationships for dynamic and rapid loading tests should only be used in soil/pile types for which they were developed and validated. Great care should be taken when using them for other soil/pile types and calibration against static testing is recommended.

6.4 Design practices to achieve higher capacities from reused foundations

Previous chapters have introduced decision models, risk, reuse factors ('R' factors, see *Chapter 3*), the importance of time, communication and confidence in material integrity. This section concentrates solely on engineering design practices to facilitate reuse. Currently, European experience of foundation reuse is limited but growing. However, there is a great deal of global experience of remediation, replacement and enhancement of foundations for structures that have approached serviceability limit state thresholds (Poulos 2004). There are direct and parallel applications of these experiences to the reuse situation.

Whether taking up reserve capacity, supplementing existing foundations or treating the ground conditions globally, the measure of success of a reuse design solution will be verifiable strain compatibility (where the inevitable differentials that occur are required to be within sensible engineering rotational and differential movement tolerances; see *Section 6.4.2*). There is variable experience of using this approach across Europe and indeed across engineering disciplines. Generally, geotechnical engineers are comfortable with the concept, many structural engineers are happier with the traditional approach when considering foundations of ultimate load/high factor of safety to limit strain, and control authorities may rely on using prescribed design values in the ground. For reuse to be successful, all of the professional team need to accept a strain compatibility approach.

Several sources of reserves and methods of supplementing foundations to achieve such compatibility have been identified and these are illustrated in Figure 6.15 (a). These consider a generic foundation situation which includes pads, rafts, slabs, piles and embedded walls shown in Figure 6.15 (b).
- *Reserves* include: surplus-capacity and under- or non-utilised elements, as discussed in *Section 6.4.1*.
- *Supplemental methods* include: global ground treatments, local element enhancement and extra foundation provision. These are discussed in *Section 6.4.2*.

6.4.1 Reserves

Potential sources of 'reserve capacity' have been identified (see, for example, *Section 6.5.1*). Certain ground (eg stiff over-consolidated clays and weak rocks) are likely to have more reserve potential than others and certain structural elements in configuration and quality may inspire more confidence. However, whatever the situation, reserve

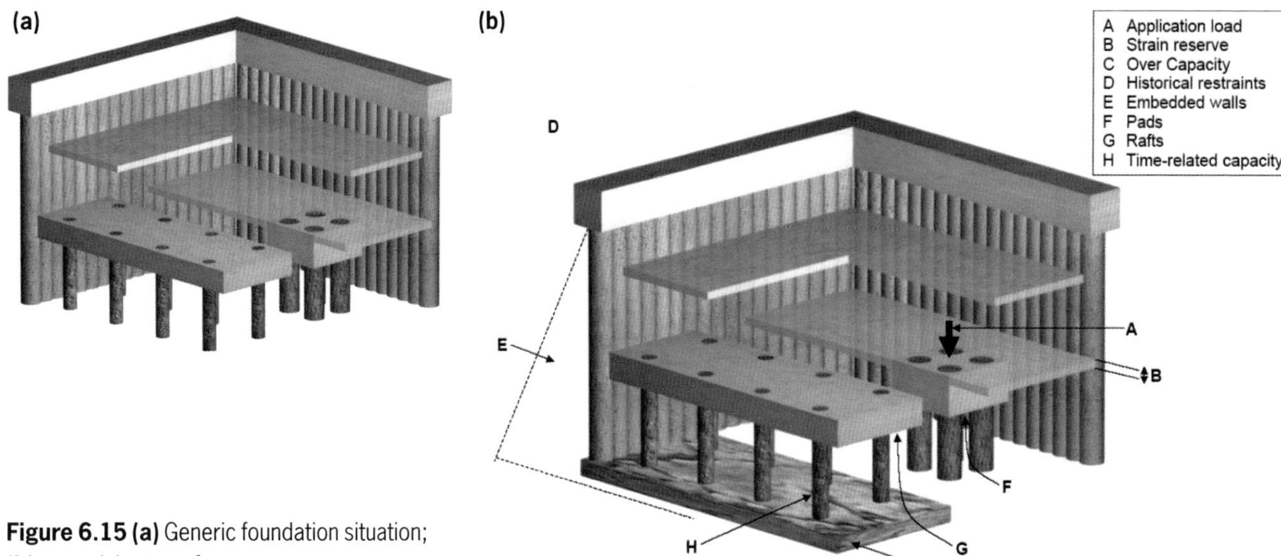

Figure 6.15 (a) Generic foundation situation; **(b)** potential areas of reserves

Box 6.7 Bankside 123 development

When a pile is designed as an element only and is predominantly a friction pile, the initially expected strain profile within it would be as illustrated. By far the majority of friction piles will be designed to this philosophy (based on moderately conservative ground strength parameters). This was the case for piles at Bankside 123.

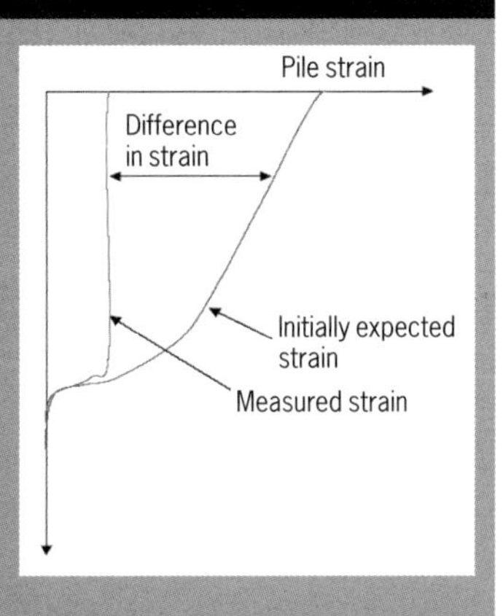

In practice at Bankside, however, most piles have a low cut-off (ie major ground removal). This means that the piles will be pulled upwards in a heaving medium. In addition, some piles have remotely located columns placed in them. These factors have resulted in the measured pile strain profile. Even after 30+ years it would be reasonable to expect a different response to load change from such piles than that assumed in the design case. This is the kind of issue that adds additional complications to reuse interpretation and analysis.

capacity cannot be assumed. It needs to be carefully researched, assessed and validated before it can be relied on for reuse (eg *Section 6.4.1.5* identifies situations where overload is not unusual).

Figure 6.15 illustrates where reserve identified in the RuFUS study may have potential in the foundation reuse situation. They include:

- *Overdesign*
  - applied load,
  - strain reserve,
  - over-capacity,
  - historical constraints.

- *Reserve capacity*
  - embedded walls,
  - pads,
  - rafts,
  - combination of elements.

- *Time-related capacity*
  - pile capacity (reload stiffness),
  - ground capacity.

6.4.1.1 Over-design: applied load

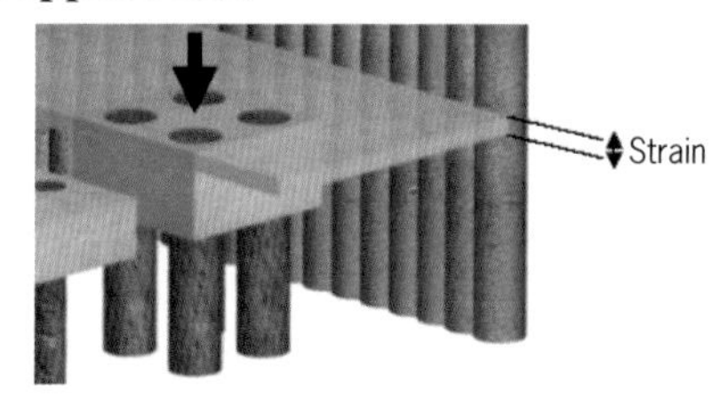

The proposed foundation reuse elements have previously been loaded by a structure. This is the equivalent of a 'field trial' and an apparent stress strain record can reasonably be assessed. Engineers can use this tool as their base parameter in terms of load and strain. Comparisons between this performance and the original predicted behaviour may indicate reserves. Limiting the load to a reduced percentage of the 'most credible' original load attempts to ensure that pile head strain does not exceed that previously experienced. The anticipated strain is accommodated within the overall foundation design (eg St John & Chow 2006).

This is possible where records are good and confidence is high regarding loading and its distribution. Confidence is also necessary in the geotechnical model and in the robustness of any analytical technique. Care must be taken when any of these features is compromised (eg when column loads are distributed to both slab and pile, rather than pile alone or when further excavation can radically affect the local strain behaviour; see Box 6.7). It should be recognised that the original calculated load (or its reduced most credible load) may be incorrect both positively and negatively. In a back-analysis situation over-optimistic predictions can soon lead to significant failings in the model.

6.4.1.2 Over-design: strain reserve

Historically, in many countries (eg UK), foundation design has been carried out using a stress-based approach with a high factor of safety (2–3) applied to a calculated ultimate capacity to produce a safe working load for the foundations. This ‘conventional design risk analysis’ (Figure 6.16) is applied to control foundation strains but the actual strain of the foundation will depend on the ground in which the foundation is installed. In Figure 6.16, foundation elements A and B would have the same safe working load using the stress-based approach. However, it is clear that element A can carry significantly more load at the same strain level than element B and still be within its elastic range, with a sensible, albeit reduced, factor of safety. This increase in load capacity, provided by using strain, rather than stress, control could provide a ‘reserve’ load capacity in a reuse situation. The approach is embedded in remediation techniques and many codes of practice accept such an approach when levels of confidence are high.

6.4.1.3 Over-design: over-capacity

Ground conditions and/or buildability considerations can lead to overprovision of capacity. Examples of this can be driving piles to a target hard layer beneath soft soils, boring piles into soft rock (buildability requires larger augers/buckets, etc. than is required for structural reasons) or where a wall founding level is deeper for seepage control than for loading requirements.

The example reported in this *Handbook* (see *Section 6.5.1*) is an ideal case study where this natural over-provision has been fully used.

6.4.1.4 Over-design: historical constraints

Local control authorities across Europe have design values and constraints regarding foundation design. These have evolved from experience and cannot be ignored. However, advances in investigation, analysis and construction knowledge and techniques make some of these rules overly restrictive, particularly if the system is viewed in strain space.

Over the last decade or so, European codes (BS EN 1997-1) have recognised that factors of safety can be varied where quality investigation and design is instigated. Reuse schemes will almost invariably require this quality of input and should be in a position to challenge overly restrictive local practice constraint.

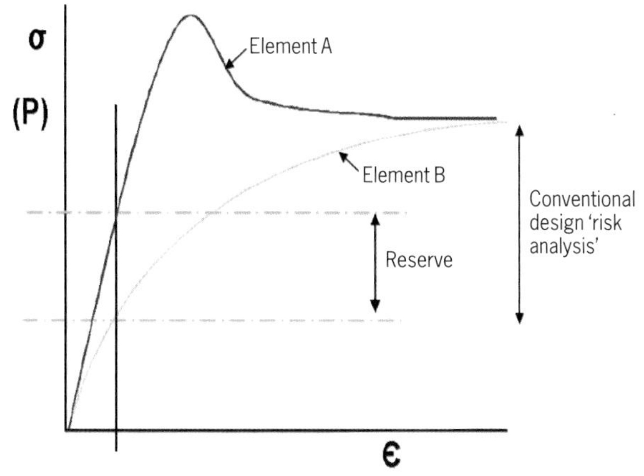

Figure 6.16 Potential strain

6.4.1.5 Reserve capacity: general issues

Figure 6.15 (b) indicates where there may be reserve capacity to take up a new load and to aid strain compatibility for the total foundation provision. In this *Handbook,* these reserves are identified as embedded walls, pads and rafts. There are many precedents for their use in foundation remediation (Poulos 2004a) and again strain is the major controlling feature.

6.4.1.6 Reserve capacity: embedded walls

Over the past few decades, European practice has varied but there are many urban situations where embedded walls are not carrying their full vertical load capacity (indeed some carry no significant vertical load at all). This applies to walls of previous decades which have potential reserve for employment in a reuse situation (Box 6.8).

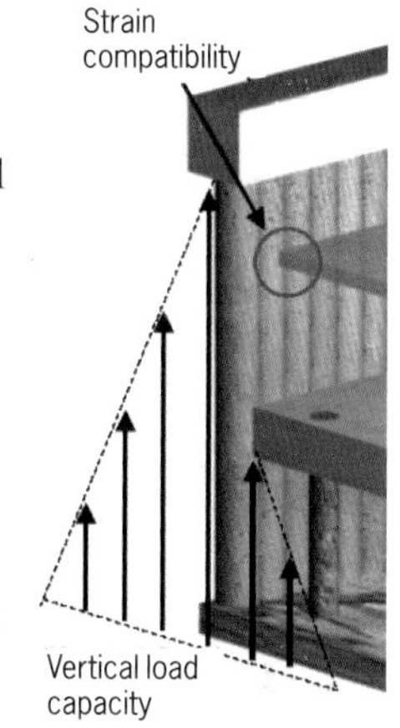

It is necessary to consider both the lateral and vertical movement of walls in foundation provision. Lateral movement is important in considering effects on adjacent buildings. This is not particular to reuse and is well-documented in literature (Burland et al 1979, Gaba et al 2003, Long 2001). Wall vertical movement is much influenced by how loads are applied and how floor and

Box 6.8 Embedded wall reserve

Embedded wall reserve may be available if further understanding of the manner in which walls carry load can be gained. The diagram, right, shows the classical means of assessment of load-carrying capacity of an embedded wall on the left of the diagram, considering only capacity below the excavation level. Within the RuFUS project, an investigation has been made into the load spread below column loads placed on an embedded wall with strain sensors placed in both capping beam and secant wall piles to determine if reserve load capacity can be identified. Potential sources include the contribution to load capacity from the retained soil as illustrated on the right of the diagram.

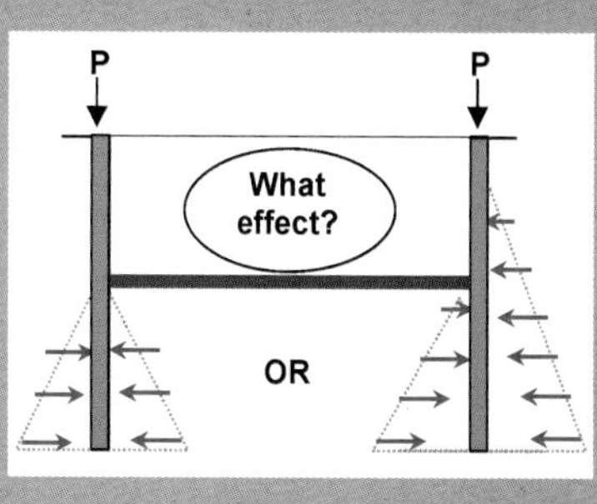

slab connections are affected. To avoid unnecessarily large stress concentrations, strain compatibility must be considered at the connection details, aiding the reuse requirement for strain compatibility across the whole substructure and highlighting the need for clear detail appreciation and robustness when reusing foundations.

Although the design of wall elements is well-understood theoretically, there is limited detailed field observation on how they actually respond to general structural loading. The RuFUS project has instrumented deep walls in a cohesive and a cohesionless subgrade in an attempt to provide a database for long-term monitoring which will clarify the soil structure–load interaction. Results from these studies and interpretation are presented in *Appendix E*. The monitoring will continue post-reuse.

Other than for simple basement situations, sophisticated and robust numerical analysis will be required to engineer sensible strain compatibility.

6.4.1.7 Reserve capacity: pads and beams

As detailed in *Section 5.3.2* there is a decision process flow chart for shallow foundations. There is a history of supplementing the load-carrying capacity of piles by using the shallower pad/beam load capacity at compatible strain levels for remedial works and this may equally be applied to reuse. The pad or beam/pile interaction scenario is a subset of the much larger set of pile/raft interaction (see *Section 6.4.1.8*).

6.4.1.8 Reserve capacity: rafts

In deep basements, piles are often connected by a slab. When this slab is ground-bearing, top face (hogging) steel is required because of the heave and water pressure effects. The same would also be true if the piles settled more than anticipated and this feature is sometimes called on in remediation schemes, ie the combined load-carrying capacity of the piles and raft at a given strain compatibility are used. This reserve may be available in considering the reuse of foundations where adequately reinforced rafts are present or where composite (thickened, integrated) slabs or rafts can be provided.

Much can be learnt from the ‘greenfield’ application of this design technique. Combined pile–raft foundations have been successfully used to:

- reduce and harmonise settlements in a cost-efficient way,
- improve the geotechnical bearing capacity of a raft foundation, and
- reduce the bending moments within the raft.

A clear model of how to apply this approach is provided in Katzenbach et al (2005).

Foundations are normally designed either as raft or pile foundation. The design of a pile foundation does not take into account the bearing reserves of the slab connected to the pile heads. The combined pile–raft is a composite foundation system. The ultimate bearing capacity and the serviceability of the whole foundation system have to be proved. In the German guidelines to combined pile–raft (CPR) construction, guidance is given on appropriate methods (Hanisch et al 2002). Given that the control is strain compatibility, in the more complex situations, it is likely that 3D finite element analysis will be required.

The bearing behaviour of the whole system depends on the behaviour of its component raft and piles. The capacity of each component depends on the strain level and the interaction effects of the adjacent components and the surrounding subsoil. For example, the contact pressure beneath a raft may cause a higher stress level in the soil between the piles causing a higher skin friction along the pile shafts. From there, the bearing behaviour of a pile which is part of a combined pile–raft foundation differs from the bearing behaviour of a comparable single pile (Figure 6.17).

To activate resisting forces of the components, settlement is required. This strain value is an important issue during the design process of a composite foundation system as it has to be compatible for all foundation elements, taking into account the stiffness of structure and soil and interaction

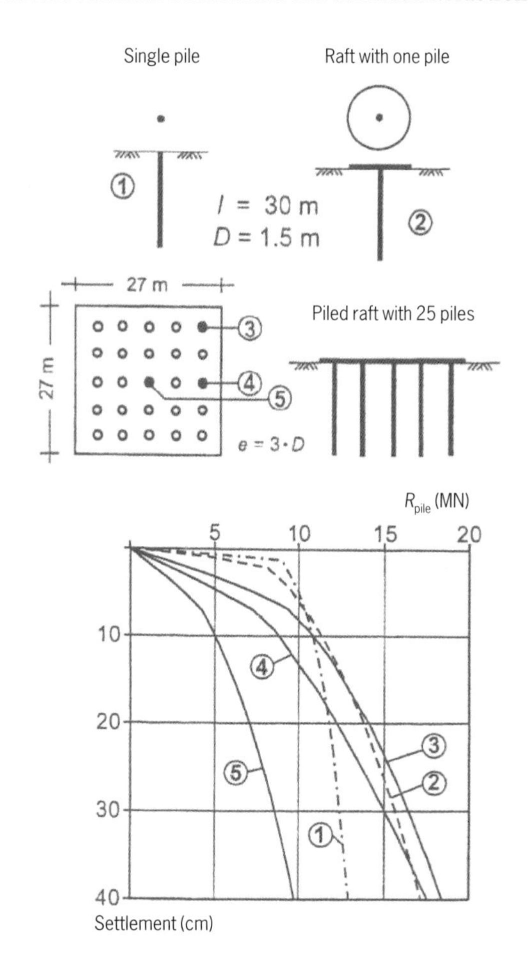

Figure 6.17 Influence of pile–pile and pile–raft interaction on load-settlement behaviour of piles (Numerical study) (Katzenbach & Moormann 2003)

effects. If a foundation element is added to an existing, perhaps still loaded foundation system, the strain required to activate the new element has to be compatible for the existing system.

It is given that, in the construction of any foundation system reliant on element interaction such as that above, workmanship and all associated requirements are to a standard that satisfies the assumed geotechnical/structural model (eg compaction of ground beneath the raft must be as per the model).

6.4.1.9 Time-related capacity

When a foundation is re-tested, the bearing capacity and performance has often changed considerably compared with the initial one just after installation. In some cases, such as driven displacement piles in sensitive clayey soils, the pile capacity may increase by more than 5 times, whereas in other cases, such as bored piles in sandy soils, there may even be a slight reduction in capacity with time. An increase in pile capacity with time is often referred to as set-up, whereas a decrease in capacity is referred to as relaxation or negative set-up.

Time-dependent phenomena are attributed to the disturbance of the soil surrounding the pile. Installation of driven displacement piles will cause significant deformation in the soil in the vicinity of the pile. For bored non-displacement piles, the local in-situ stress conditions will be affected by the piling method. While drilling, stress release will take place. A small plasticized annular zone around the pile may also result which will subsequently partly be recompressed by the concrete pressure.

When loading the pile, the loads transmitted by shaft friction are transmitted through these disturbed zones around the pile. With time, 'healing' (referred to as ageing) may take place which affects the shaft friction performance.

The aim of the next sections is to present an overview of what is considered to be current knowledge in the field of long-term behaviour of pile foundations.

Driven piles in cohesive piles

The fact that the axial capacity of a driven pile in clay may change over time has been documented for more than a century. Long et al (1999) presented a database of information available in the literature: Figure 6.18 shows results for both static and dynamic tests on a range of piles in clay from their work.

Komurka et al (2003) suggested there were three phases in the 'set-up' mechanism for piles in cohesive soils. In the first two, pore pressures generated during driving dissipate at rates depending on the local soil conditions the consolidation period in Figure 6.18. Capacity calculated during driving may include significant effects of pore pressure, and will be different from that found in tests carried out soon after driving. Pore pressure may take months to dissipate. In the third phase (ageing period in Figure 6.18), dissipation of pore pressures has been completed and change in capacity continues at nominally constant effective stress and generally

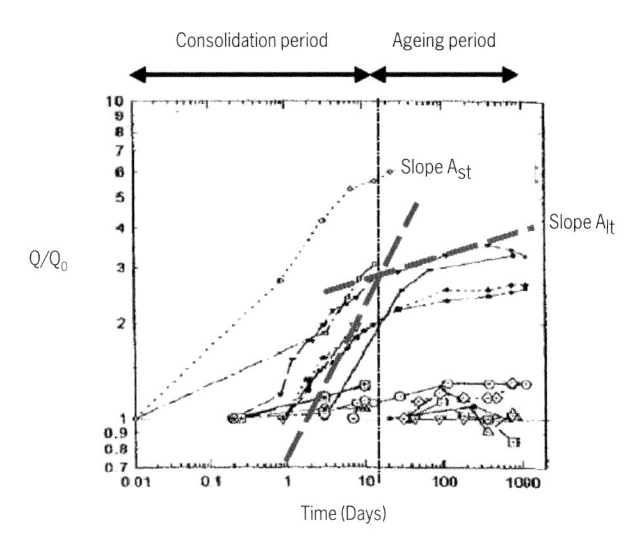

Figure 6.18 Examples of long-term performance of driven piles in clays

at a slower rate (A_{lt}) than during the consolidation period (A_{st}).

Although the mechanism behind the ageing effect in clay is still not totally understood, it appears to have an important and long-lasting effect on the pile capacity. The significance of this stage was shown by Powell et al (2003) who reported the results of a series of load tests on steel driven and jacked piles in clay. The piles were retested more than 20 years after their installation. The results are reproduced on Figure 6.19, in the form of normalised capacity (capacity at time, t/capacity in initial testing) against time and show significant long-term changes.

Driven piles in granular soils

In dense or very dense sands, high resistance may be observed during driving due to negative pore water pressures generated. With dissipation of the negative pore pressures effective stresses decrease, resulting in a reduction in pile capacity. In loose sands, the opposite is more likely and pile capacity may increase after dissipation of excess pore pressures. But, in contrast to clayey soils, dissipation of positive or negative pore pressures takes place rapidly (generally less than 1 hour).

After dissipation, a further set-up is usually observed. Many academic, field and laboratory studies have been performed to identify the mechanisms behind this stiffening and strengthening effect (Chow et al 1997, 1998). The main mechanisms put forward in the literature are the creep behaviour of sand particles, leading to the change in the stress distribution around a pile, and an ageing effect, leading to an improvement in the mechanical soil characteristics at constant effective stress. Aging is attributed to cementation at surface contacts between sand particles after deposition or densification (Lee 1977).

The results of a large series of pile load tests in sands have been reported by Long et al (1999). The axial capacity is shown in several cases to increase by a factor of approximately two and continues over a long period, up to 500 days after driving.

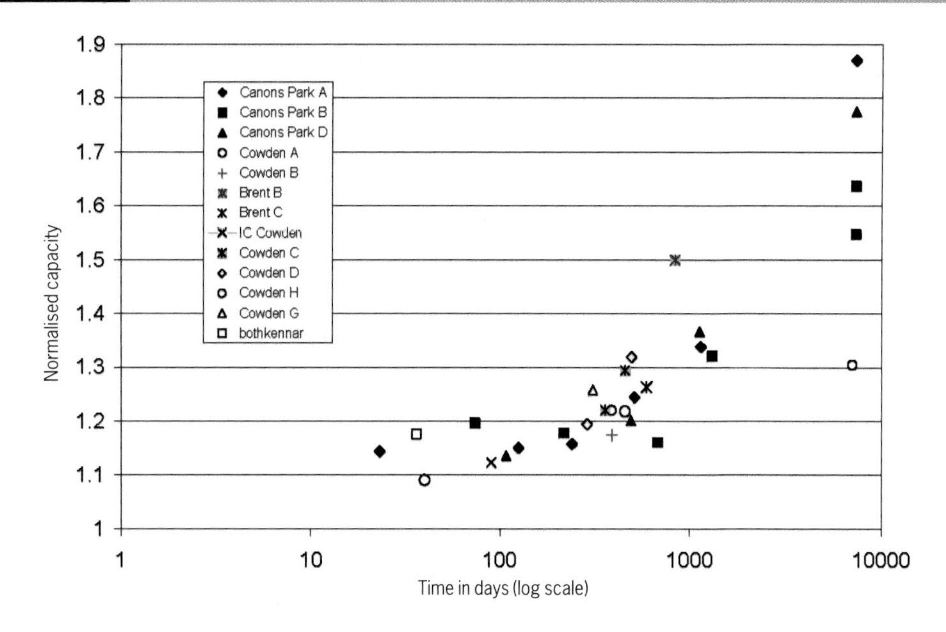

Figure 6.19 Normalised capacity against time

Models to predict the long-term behaviour for driven piles
Several models have been developed to predict the set-up for driven piles. The model proposed by Skov & Denver (1988) predicts a semi-logarithmic capacity increase with time: where:

$$\frac{Q_t}{Q_0} = 1 + A\log\left(\frac{t}{t_0}\right)$$

Q_t = axial capacity at time, t, after driving,
Q_0 = axial capacity at time, t_0,
A = dimensionless set-up factor, depending on soil type.

Although this model is commonly used, it presents some drawbacks as follows.

- The model defines no upper limit for capacity, although in practice the rate of increase will decrease.
- The set-up factor, A, depends on the value considered for t_0, the time of initial testing. In many publications, the reference capacity of the driven piles is deduced from the driving parameters, and is referred to as Q_{EOD} (where EOD means 'end of driving'). The reference time in that case is taken as $t_0 = 0.01$ days. Large changes in pile performance may be noted during the first minutes after driving, so to minimise the influence of this initial effect, a longer period after installation is preferable for the reference time. In practice, the following figures are recommended for this model: $t_0 = 0.5$ days for sandy soils and $t_0 = 1$ day for cohesive and mixed soils.

Other models have been developed to overcome the shortcomings of the Skov & Denver model that require more detailed information (which is rarely available) on the pile tests.

Bored non-displacement piles
Unlike driven piles, few case studies are available that investigate the set-up of bored piles. They generally report an increase of the bearing capacity with time, but to a much lesser extent than that observed for driven displacement piles.

An extensive and detailed case study has been reported by Bullock (2003) in 'a study of the set-up behaviour of drilled shafts'. It provides the results of the re-testing of five piles previously loaded in 1996, ie 6.5 years earlier. The pile lengths varied from 26 m to 30 m with diameters from 1.5 m to 2.15 m. The piles were heavily instrumented with strain gauges providing a shaft load profile from which it is possible to estimate a set-up factor for shaft friction in different soil types: clay, sandy clay, sand and limestone. Table 6.5 reproduces some key figures resulting from this study.

The average set-up factors in the different soil layers are very similar and are small compared with the published set-up factors for driven piles. However, the figures in the table are average values and the variation was large. Some piles even showed negative set-up (or relaxation).

As part of the RuFUS project, Powell & Skinner (2006) report the first stages of a study of bored piles in heavily over-consolidated clay. Comparing results of testing 'virgin' piles over 3.5 years they show capacity increases approaching 30% compared with tests 2.5 months after installation. This is equivalent to an A value of 0.25, based on $t_0 = 1$ day. They also compared results on re-tested piles and in fact show a reduction in capacity compared with initial testing. It seems

Table 6. 5 Example of shaft friction set-up factors for drilled shafts (Data from case histories reported by Bullock 2003)

Soil type	Number of data	Average shaft friction (kpa) At t_0 = 1 day	Average shaft friction (kpa) At t = 2000 days	Average set-up factor, A
Clay	7	10	14	0.12
Sand	12	49	65	0.05
Clay–sand	6	100	125	0.08
Limestone	6	470	547	0.10

Box 6.9 Ageing

Ageing need not always produce capacity increase. For example, many major European cities are experiencing rising groundwater which can reduce both end-bearing and skin friction components of pile capacity, and affect strain behaviour. This illustrates the effort necessary in assessing global conditions when dealing with the ground and the care required in back analysis of apparently stable conditions. It adds another dimension to foundation reuse.

likely that some period after testing is needed for the soil to 'heal' after testing, similar to driven piles.

6.4.2 Supplemental methods

The reserve capacity section (*Section 6.4.1*) has illustrated some of the issues that can be addressed/used to aid reuse. Figure 6.20 illustrates where supplemental methods can be employed. These include:

- global scheme modifications below the structure footprint, usually to the ground in such a way that its properties are enhanced and made more homogenous in its reaction to load,
- local elemental improvements to structural elements such as piles or barrettes to improve local behaviour and to carry particular loading or reduce local movement,
- provision of additional elements to supplement or complement existing features.

While these issues are separated for the purposes of this *Handbook*, there is no reason why methods from *Sections 6.4.1* and *6.4.2* cannot be mixed. Again, it should be stressed that strain at the ground–structure interface (head of a pile/top of raft/junction slab and walls) will generally be the overall controlling feature for reuse purposes.

In the reuse situation, access will seldom be 'greenfield' and will often have several restrictions (see Case studies, *Sections 6.5.1* and *6.5.2*). As illustrated, headroom can have a particular influence on available options and the RuFUS study has tended to concentrate on solutions that use lightweight, low headroom plant to facilitate provision of any supplemental methods. However, as some of the significant case studies illustrate, there are situations where headroom has not been restricted.

6.4.2.1 Global (ground) modification

Classical ground modification/improvement techniques include strain control by: effective stress change by controlling groundwater (pumping, electrokinetics); stiffness enhancement (compaction grouting, jet grouting, chemical grouting, compensation grouting, mix in place, electrophoresis, ground baking, ground freezing) and designed soil removal (Burland et al 2003, Poulos 2003, 2004a).

These have been used in many situations for ground remediation works. Although the RuFUS survey did not identify many direct case studies, there is no reason why these might not be considered for the reuse situation. The likely target is a 'raft-like' foundation support that will be as homogeneous as is practicable to deliver strain compatibility across the structures footprint. The stabilisation of the Pisa tower and the jet grouting on the west access gallery of the Grand Palais, Paris, (see Case study, *Section 6.5.2*) illustrate where a ground improvement method has been used to ensure strain compatibility for a major reuse situation.

6.4.2.2 Pile improvement techniques: general issues

Pile capacity improvement techniques can be used to enhance the geotechnical bearing capacity of an existing local pile foundation, provided the structural capacity of the piles is able to accept the new loads.

Figure 6.21 indicates schematically the principal zones involved with the load transfer from a pile into the resistant soil layers. These zones can be targeted when designing capacity enhancement.

Several pile improvement techniques have already been used in the past, mainly on bored piles. Their application

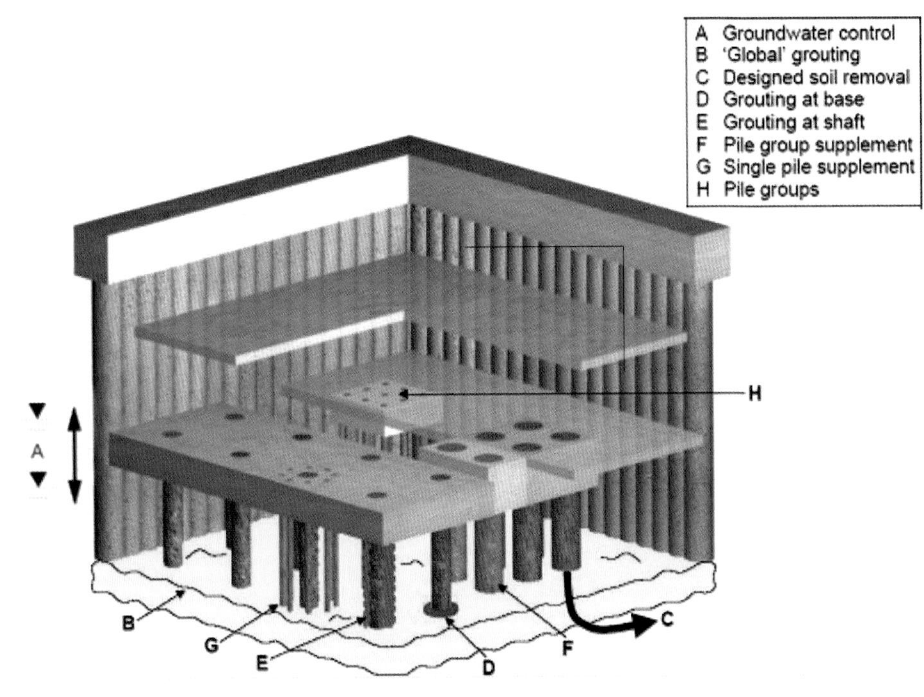

Figure 6.20 Possible supplemental methods to enhance the foundation capacity

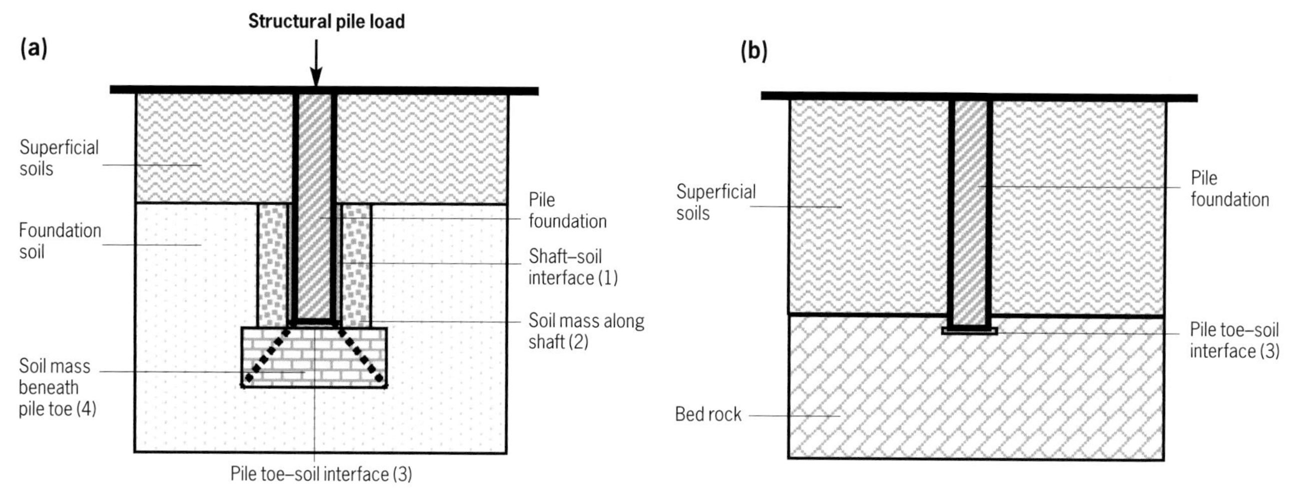

Figure 6.21 Potential application areas for pile improvements: **(a)** pile foundation in soils, **(b)** pile foundation on bed rock

Table 6.6 Potential application areas for pile improvement techniques for bored piles and their application possibilities for RuFUS projects

	Lateral friction		End-bearing	
	1 Pile shaft–soil interface	2 Soil mass around pile shaft	3 Pile toe–soil interface	4 Sold mass beneath pile toe
Possible shortcomings	Poor contact (low mechanical characteristics, reduced soil pressures)	Decompressed soil volume	Remoulded and/or decompressed soil	Poor mechanical characteristics
Possible consequences	Reduced lateral friction	Reduced stiffness	Reduced stiffness	Reduced end-bearing

Table 6.7 Pile improvement techniques for bored piles and their application possibilities for RuFUS projects

		Lateral friction		End-bearing	
Treatment	**Treatment principle**	1 Pile shaft–soil interface	2 Soil mass around pile shaft	3 Pile toe–soil interface	4 Sold mass beneath pile toe
Jet grouting	Creation of a soil–concrete (mixture of soil and cement) of controlled dimensions and mechanical characteristics				Well suited Transforms the soil mass beneath the pile in a soil–concrete
Shaft grouting	Injection of a rigid cement slurry near the shaft–shaft surface constitutes a preferential flow path for the slurry	Well suited Improved mechanical characteristics: improved contact pressure			
Permeation grouting	Injection of a high penetration grout to improve the mechanical characteristics of the soil		Possible Improved soil characteristics: can be efficient in coarse-grained soils		Possible Improved soil characteristics: efficient in coarse-grained soils
Compaction grouting	A stiff mortar is grouted in the soil to form mortar columns and to densify the surrounding soil		Possible Re-compression of the soil mass, but reduced efficiency		
Base grouting	Injection of a rigid cement slurry at the contact			Possible Improved stiffness, but requires grouting point at pile toe	

potential on RuFUS projects is highlighted in Tables 6.6 and 6.7. Two improvement techniques, shaft grouting and jet grouting at the base, seem particularly appropriate.

6.4.2.3 Pile improvement techniques: shaft grouting

By using appropriate grouting techniques, it is possible to create a skin around the shaft of a pile, with a thickness of several (5–15) millimetres. The benefits of such a grouted skin are that it improves the effective horizontal soil pressure and the mechanical bond between pile surface and soil.

Shaft friction improvement factors from 1.5 to 7 have been recorded in case histories. Grout type, grout quantity and grouting pressures are important parameters and must be carefully designed and monitored. However, as illustrated in Figure 6.22, based on the RuFUS database, experience is not yet sufficient to draw reliable relationships between these parameters and the achieved pile capacity increase.

French standards [DTU 13.2, Fascicule n°62 – Titre V (AFNOR 1992)] give design indications for the load-carrying capacity of grouted piles. Table 6.8 indicates the lateral friction improvement factor, which can be calculated from these documents. Improvement factors range between 2 and 5, ie in the same range as the field observations.

6.4.2.4 Pile improvement techniques: jet grouting

During the jet grouting process, the natural soil structure is eroded by high pressure water or grout jets and the soil residue is mixed with the grout, partly mixed or removed and replaced with fresh cementitious grout. It is possible to create

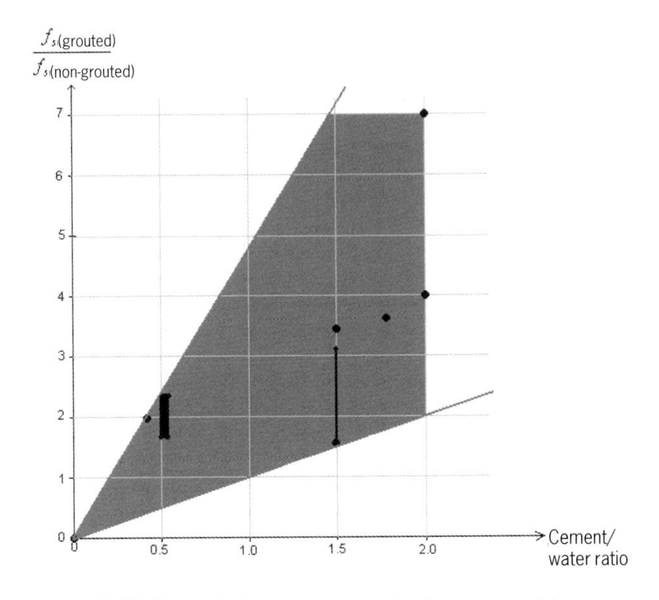

Figure 6.22 Shaft friction improvement ratio vs cement/water ratio (Derived from literature)

Table 6.8 Shaft grouting: average shaft friction improvement factor (According to French DTU 13.2 and Fascicule n°62)	Clay, Silt	Sand, Gravels
Bored	2–4	1.5–4
Bored with bentonite slurry	2–5.5	2.2–5.5
Bored with casing	2–5.5	2.2–5.5

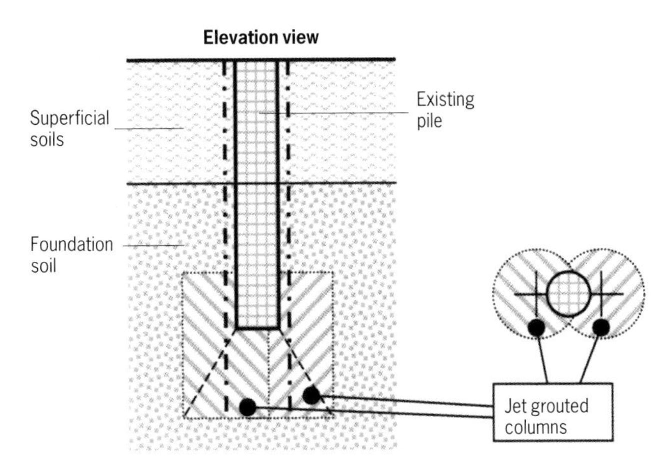

Figure 6.23 Jet grouting: possible treatment scheme for base capacity enhancement

soil–mortar columns of controlled dimensions beneath an existing foundation, provided appropriate design, execution and monitoring procedures are used. Such grout columns can be designed to deepen and enlarge existing pile toe foundations (Figure 6.23).

The technique is versatile and all soils, from weak clays to dense sands, can be treated, although with increasing compactness or stiffness of the soil, the efficiency (ie penetration radius) of the jet grouting will decrease. In practice, the application possibilities, within the context of reuse projects, will most likely focus on granular non-cemented soils.

Classical geotechnical design formula can be used to estimate the improvement of the end-bearing capacity. Depending on the number and size of the columns, the improvement can be significant. For the two case histories (Bustamante et al 1996, Ravaschio et al 2000) where this method has been used, improvement factors of 2 and 9 have been reported.

An important limitation with jet grouting is that the soil beneath the pile toe will temporarily be transformed into a liquid soil–mortar with no strength, until the cement has set sufficiently. This means that, for a short time, the pile will not have any toe resistance so improvement with jet grouting techniques can only be considered for those cases where the pile toe will be temporarily in a non-loaded condition.

6.4.2.5 Element supplement

It would be fortuitous if the new build load action points coincide with the pile or pile group reaction points (*Geometric compatibility* as defined in *Section 5.2.3*). The majority of cases will require supplemental elements which will comprise either load transfer beams, slabs and pads or embedded elements which directly carry the load or influence the reaction centroid geometry. Early and close involvement with the structural engineer to effect load transfer at or above the ground can be cost-effective and should always be investigated. This section, reviews the contribution that can be expected from ground embedded load-carrying elements.

Figure 6.24 identifies some supplemental configurations that are being/have been employed. In all cases, it is strain compatibility that is the reuse control driver. This is a complex process and there will be virtually infinite configurations dependent on the circumstances. Figure 6.24 illustrates how many configurations can be built from two simple elements. For this section, two separate supplement cases have been identified, the first (1 a & b) where the supplemental works are considered discrete units distinct from existing piles, the second (2 a & b) where the new piles supplement the existing piles both locally and also globally.

(1a) The simplest application of this is in an unrestricted headroom situation where a mirror reflection of an existing pile can be provided at a spacing at which the centroid of load application and reaction coincide. Even this simple case illustrates the level of confidence necessary in the likely performance of the existing pile.

(1b) Often in the re-use situation headroom is restricted and only small, lightweight, low headroom, specialist plant can effect access. In such a situation, it may be possible to mimic the large existing pile strain behaviour by the provision of small diameter pile groups in place of a single large diameter pile.

(2a) This shows examples of where smaller diameter piles are used locally around an existing pile to influence the strain behaviour at the head of the pile group. Poulos (2004a, b) shows examples of this. Load transfer mechanisms between pile types are more complex in this situation and the use of controlled stiffness inserts (CSIs; see Box 6.10) may increase confidence levels in ensuring compatible behaviour (Poulos 2004a).

(2b) The interface between piles and rafts is important. Sometimes, piles are required to carry specific loads and at other times they are used as settlement reducers. Mixed raft/pile solutions are often required for reuse situations. Almost without exception, they will need advanced numerical analysis to assess relative movements sensibly (Vaziri 2005), and increasingly 3D analysis will be called for. In stiff sub-soils with relatively small strains, it may be possible to use semi-empirical assessments and simplified load distributions (Box 6.11) at initial feasibility/cost stages.

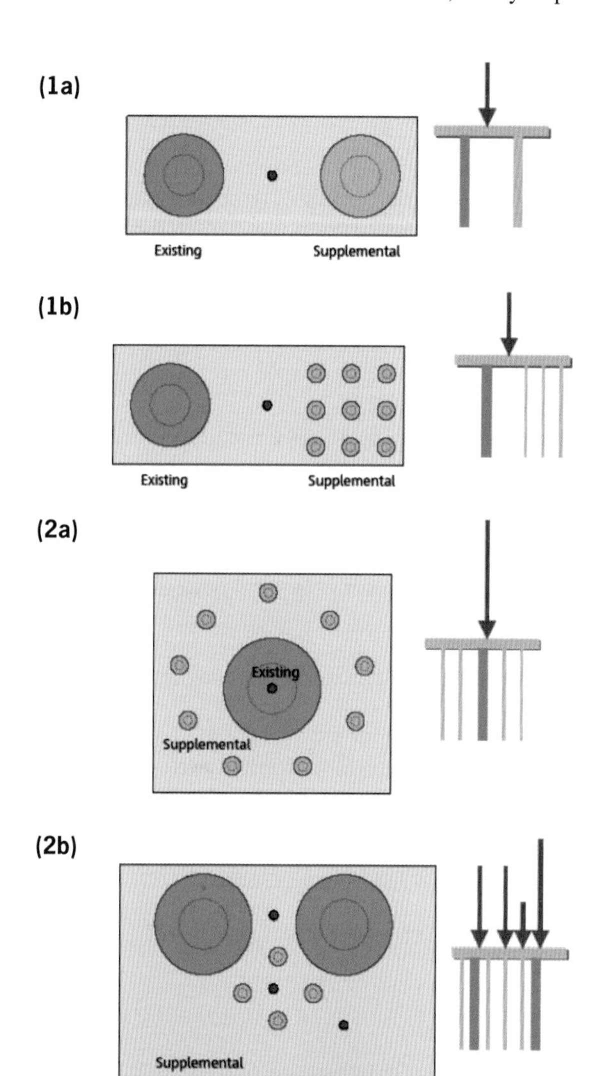

Figure 6.24 Element supplement

Box 6.10 Controlled stiffness inserts

The use of controlled stiffness inserts (CSI) is described by Poulos (2004a, b). These inserts which may be neoprene or similar semi-compressible material serve to decrease the pile head stiffness in a controlled manner (hence controlled stiffness inserts). They were used at the heads of remedial pre-bored H piles in a housing project in Hong Kong.

Such controls have been under consideration for many years but this appears to be one of the first reported uses. The advent of advanced numerical analysis, the ability to involve 3D analysis and closer cooperation between the structural and geotechnical engineer make possible the acceptance of CSI techniques. As reuse of foundations increases and strain control is recognised as being the important parameter, CSI techniques may have a significant role in the future.

Box 6.11 Feasibility studies (Klar et al 2004)

For feasibility studies, Klar et al (2004) have developed a fast computational algorithm which reduces the arithmetic operation significantly. The method eliminates the need for full matrices and the power of the procedure is even greater when considering the optimisation problem and statistical analysis. The authors are not suggesting that such a tool will replace finite element (FE) analysis, but it can be used in sensitivity studies to help in choosing the cases for which the more detailed FE analysis should be conducted.

6.5 Case studies

The key issues of verification in a reuse situation are the establishment of integrity and compatibility. The relative cost of such establishment will often dictate reuse viability. The following examples illustrate the interplay between these fundamental elements and their influence on the reuse decision.

6.5.1 Belgrade Plaza, Coventry, UK

4-storey car park being raised a further 4 storeys on the same footprint. Foundations generally 3 piles per cap. Piles in weak rock, therefore, reserve available.

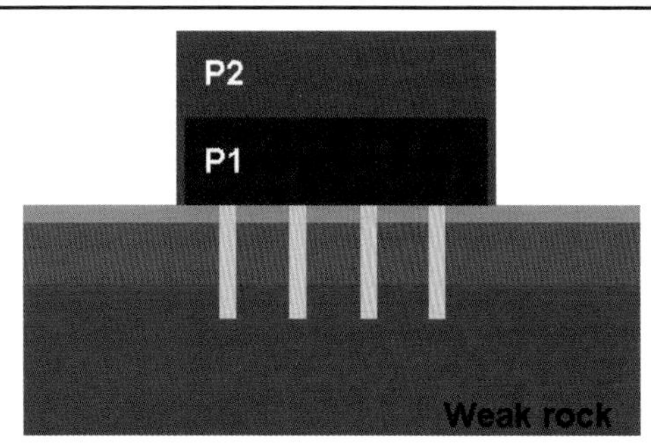

Assessment

Desk study

City with well-understood ground conditions. Strict building control. Built 26 years ago. For buildability reasons, existing piles were a minimum of 600 mm diameter with minimum penetration into weak rock ensuring a large load capacity reserve. The extra load could be channelled through the existing columns. The alternative proposal requires new foundations, new columns, slab breakthrough albeit on the same footprint. Construction records and original material and pile tests available. Piling contractor prepared to sell warranty.

See Sections 6.2, 4.4

Investigation

Investigation comprised pile coring through to weak rock. Rock and concrete crushing tests. Confidence levels high, therefore, investigation of ground and material limited to coring 4 piles from a total of over 200 for integrity proofing of the concrete and underlying rock. Eccentric loading no concern.

See Section 6.3.2

Performance

Confidence in previous pile test results (2 no.). Same pile company proposing foundation reuse.

See Section 6.3.4

Design

Compatibility

Strain in such ground is small so achieving compatibility is not difficult. Positional compatibility of loading.

See Sections 6.4, 5.2.3

Reserves

Reserves proven because of over-capacity arising from buildability requirements.

See Section 6.4.1.3

Supplemental methods

New minipiles provided to enhance confidence levels. Rock penetration chosen to ensure strain compatibility. Issue is bond in existing cap.

See Section 6.4.2.5

Verification

Professional team accept previous test results from other sites on existing cap bond. Local monitoring of pile using strain sensors and global monitoring of foundation behaviour using precise levelling (fallback position is introduction of extra minipiles should any trigger levels be breached).

See Chapters 4, 5, 7, Section 6.3.2

Costs

Effect on superstructure costs

Reduces superstructure costs as demolition and additional columns not required.

See Section 2.5.2

Effect on foundation costs

Reduced foundation costs as only supplemental minipiles required.

Effect on verification costs

Low costs for verification works and existing pile warranty.

Outcome

Successful reuse of existing foundations with high level of confidence in performance and clear verification of results. Excellent model for design of foundations for reuse.

See Section 7.2 and Tester & Fernie (2006)

6.5.2 Grand Palais, Paris, France

Rehabilitation of the Grand Palace, a landmark building in Paris, built between 1897 and 1900. Works included the provision of a 12 m deep underground space for future development. The southern part of the building is located on an old river bed, filled with clayey deposits. The old foundation, consisting of 3200 oak piles, had deteriorated and needed to be replaced. In the northern part, the building is founded on strip foundations, which had to be supplemented to achieve a compatible stiffness with the new foundations in the southern part.

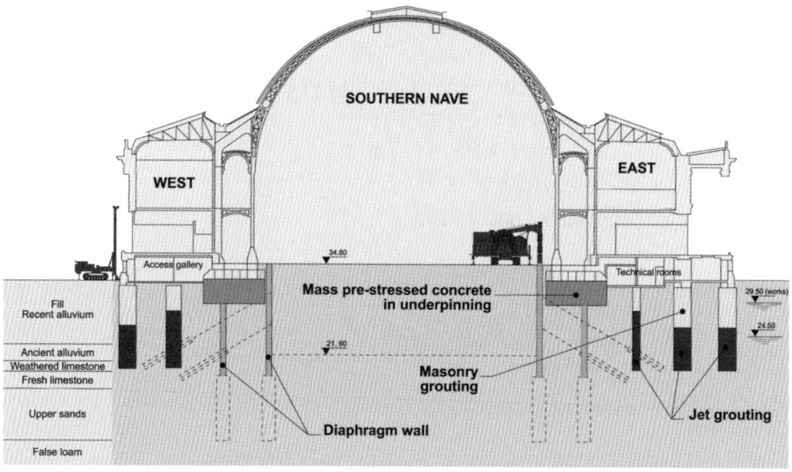

Assessment

Desk study

Detailed archives of the design and construction period are still available, as are monitoring records from 1910 (see below).

See Section 6.2.2

Investigation

To enable the renovation project, a detailed soil investigation was implemented over the full building area, with pressuremeter tests down into the bedrock, and investigation of the pile integrity in the southern part.

See Section 6.3.3

Performance

Flow management works on the Seine led to a drop in groundwater levels. Because the building was located near the Seine, this inevitably affected the integrity of the piles, causing differential settlements of between 100 and 120 mm. Following flooding of the area in 1910, the building began to show signs of structural problems. From 1930 onwards, a systematic cataloguing of problems was undertaken. In 1993, a rivet falling from the building triggered the renovation programme. The severe deterioration of the top of the wooden piles in the southern part of the building prevented the piles from being integrated into the new foundation scheme. In the northern part, where the building is founded on strip footings, the settlement behaviour was satisfactory.

See Sections 6.3.4, 4.3.1

Design

Compatibility

Despite sufficient mechanical bearing capacity of the strip foundation in the northern part, it was deemed necessary to stiffen these strip foundations to achieve a strain behaviour compatible to that of the new foundations in the southern part.

See Sections 3.2, 6.4.1.8

Reserves

No hidden reserves have been considered in the design.

See Section 6.4

Supplemental methods

In the southern part, the wooden pile foundation was fully replaced by RC diaphragm walls down to the limestone. The wall was designed to allow for future 12-metre excavation. In the northern part, the strip foundation was stiffened by jet grouted columns down to the limestone.

See Sections 6.4.2.5, 6.4.2.4

Verification

Classical geotechnical design, considering strength and strain behaviour of the different foundation elements.

Costs

Effect on superstructure costs

No alternative solution was considered.

See Section 2.5.2

Effect on foundation costs

None.

Effect on verification costs

None.

Outcome

Successful rehabilitation of a prestigious but complex building by controlling strain to minimise deferential settlements locally and globally.

6.5.3 Belgrave House, London, UK

6-storey steel structure with single basement replacing existing 7-storey concrete frame structure with single-storey basement. The foundations were mainly under-ream piles founded a minimum depth of 16 m into London Clay (approx. –19 m OD), under-ream diameters varying from 2.3 to 2.8 m with some straight-shafted piles without under-reams.

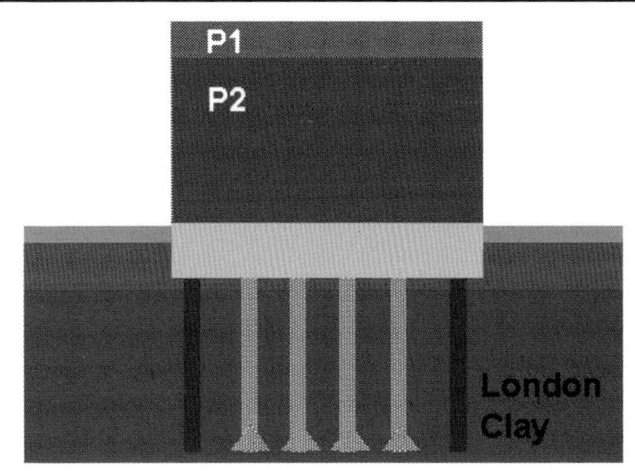

Assessment

Desk study

City with well-understood ground conditions, strict building control, good trading practices, reputable consultant and piling contractor both still trading. No as-built records or piling close-out reports available. However, design drawings (circa 1972–1978) were obtained from the consultant. Back-analysis using a moderately conservative strength profile of the London Clay predicted safe working loads, 7–10% greater than the original.

See Sections 6.2, 6.3.2

Investigation

Confidence in pile position/diameter at post-demolition stage. Knowledge of behaviour of under-ream piles in similar ground conditions. A check was carried out on the pile heads following demolition to confirm structural integrity and pile toe level of 5% of piles. Visual inspection (including microscopic inspection) and laboratory tests on cored pile concrete on 5% of piles confirmed that there was no material deterioration.

See Sections 6.3.2, 6.3.3

Performance

NDT confirmed pile lengths. Monitoring of some of the pile heads during deconstruction showed no unexpected strain behaviour.

See Sections 6.3.4.1, 6.3.3

Design

Compatibility

Although the overall loading did not significantly increase, total geometric compatibility was incomplete. Sophisticated numerical analysis was carried out to ensure efficient load transfer between load action points and the piles (existing and some new straight-shafted).

See Sections 5.2.3, 6.4

Reserves

Reserve not relied on. However, overall load slightly less than previous load (confidence levels high regarding past load history).

See Section 6.4

Supplemental methods

Deeper straight-shafted piles where load concentrations were incompatible with existing pile arrangement. Piles sized to ensure strain compatibility. This was easily achieved as the strain difference was small.

See Sections 6.4, 6.4.2

Verification

A strategy for the reuse of the piles was developed at the outset to address the design risks and successfully followed through the demolition phase to confirm their suitability for reuse. An alternative foundation design scheme was drawn up in the event of the existing piles not being reused. The client, their insurers and the consultant evolved a risk management process for conditions of reuse.

See Chapters 6 and 7

Costs

Effect on superstructure costs

Neutral as original concept carried through.

See Section 2.5.2

Effect on foundation costs

Cash positive since transfer slab and new pile cost significantly less than total new build.

Effect on verification costs

Higher but see above.

Outcome

Successful and confident reuse of existing foundations with reduction in construction programme and lower foundation cost.

See also Vaziri & Windle (2006 a,b)

6.6 Key points

The investigation and design of foundations for reuse adds complexity to the design situation and requires a significant amount of interaction between the designer, constructor and other parties to the development scheme.

- The reuse of foundations situation requires the continual viability appraisal between the assessment and design processes as opposed to the near-linear process for new foundations (see Figure 6.1).
- The desk study and initial investigation phase allows: not only the physical parameters regarding the ground and the components that are placed in the ground to be categorised/defined/established/inferred but it also gives an opportunity to set in context the other important features in the foundation provision including quality control, social context, historical context (capabilities and limitations) and risk attitudes.
- NDT has a large part to play in reuse foundation assessment. No NDT method is complete, all have to be calibrated to other indices (eg physical testing, other NDT, predictions, desk study knowledge). NDT-CE requires sophisticated and specialist interpretation.
- Movement compatibility is a major consideration in foundation reuse, more so than ultimate load capacity. This changes the emphasis from wishing to obtain information on ultimate load capacity to the need to understand the load (strain cycle). Close observation of strain at the deconstruction or load-testing stages becomes more important. NDTs that provide load capacity information only should be viewed as only another index.
- Reserve capacity may exist (although this cannot be expected) and rigorous assessment procedures must be followed before using such reserve. Reserves identified in this *Handbook* include over-design in load, strain and provision, reserve capacity in under-utilisation of structural elements and its time-related capacity in ground–structure interaction.
- Although reuse experience is limited, ground remediation experience is worldwide. Supplemental methods have been identified, including ground modification, pile improvement and element addition techniques. In the urban situation, it is likely that access and headroom restrictions will mean that light and low headroom plant will often be required.
- Foundation reuse lacks a significant history so requires appropriate verification methods at all its levels (material, model and analytical confidence and performance). Observation, instrumentation and monitoring are key issues to the reuse engineer.

6.7 References

AFNOR. DTU 13.2, Règles techniques de conception et de calcul des fondations des ouvrages de Génie Civil: Fascicule n°62. Titre V. 1992

AFNOR. *NF P 94-160-3: Auscultation d'un 'el'ement de fondation. Partie 3 : M'ethode sismique parallèle (MSP).* Paris, Assosciation francaise de normalisation, 1993

BRE. *Site investigation for low-rise building: desk studies.* Digest 318. Bracknell, IHS BRE Press, 1987

BRE. *Site investigation for low-rise building: the walk-over survey.* Digest 348. Bracknell, IHS BRE Press, 1989

British Standards Institution. BS EN 206-1: 2000 Concrete. Specification, performance, production and conformity

British Standards Institution. BS EN 1997-1: 2004 *Eurocode 7. Geotechnical design. General rules*

Bullock PJ. A study of the setup behavior of drilled shafts. Final Report Contract #BC-354, work unit #RPWO 32, University of Florida, 2003. 462 pp

Burland JB, Simpson B & St John HD. Movements around excavations in London Clay. Invited National Paper. *Proceedings 7th European Conference on Soil Mechanics and Foundation Engineering,* Brighton, 1979. Vol 1, pp 13–29

Burland JB, Jamiolkowski M & Viggiani C. The stabilisation of the Leaning Tower of Pisa. *Soils and Foundations* 2003: **43** (5): 63–80

Bustamante M, Gianeselli L & Thiriat D. Strengthening of a viaduct built at the beginning of the 20th century by jet-grouting. In: Viaggiani C (ed) *Geotechnical engineering for the preservation of monuments and historical sites. Proceedings International Symposium,* Napoli, Italy, 3–4 October 1996, pp 543–550

Butcher AP. The detection of pile geometry using geophysics. In: Butcher AP, Powell JJM & Skinner H (eds). *Reuse of Foundations for Urban Sites: Proceedings of International Conference,* BRE, Watford, 19–20 October 2006. EP 73. Bracknell, IHS BRE Press, 2006. pp 87–94

Butcher AP, Campanella RG, Kaynia AM & Massarsch KR. Seismic cone downhole procedure to measure shear wave velocity: a guideline prepared by ISSMGE TC10: Geophysical Testing in Geotechnical Engineering. *Proceedings of XVI ICSMGE,* Osaka, Japan, 2005

Butcher AP, Powell JJM & Skinner HD (eds). *Reuse of Foundations for Urban Sites: Proceedings of International Conference,* BRE, Watford, 19–20 October 2006. EP 73. Bracknell, IHS BRE Press, 2006a

Butcher AP, Powell JJM & Skinner HD. Stonebridge Park: a demolition case-study. In: Butcher AP, Powell JJM & Skinner H (eds). *Reuse of Foundations for Urban Sites: Proceedings of International Conference,* BRE, Watford, 19–20 October 2006. EP 73. Bracknell, IHS BRE Press, 2006b. pp 321–330

Chow FC, Jardine RJ, Brucy F & Nauroy JF. Effects of time on capacity of pipe piles in dense marine sand. *Journal of Geotechnical and Environmental Engineering* 1998: **124** (3): 254–264

Chow FC, Jardine RJ, Nauroy JF, Brucy F. Time-related increases in the shaft capacities of driven piles in sand. *Géotechnique* 1997: **47** (2): 353–361

Deutsche Gesellschaft für Zerstörungsfreie Prüfung (DGfZP) [The German Society for Non-destructive Testing]. *DGZfP-Fachausschuss für Zerstörungsfreie Prüfung im Bauwesen (AB) Unterausschuss Radar* [Technical Committee NDT in Civil Engineering], *Merkblatt über das Radarverfahren zur Zerstörungsfreien Prüfung im Bauwesen* [Guideline for the radar method in NDT-CE] (Merkblatt B 10), 2001

Deutsche Gesellschaft für Zerstörungsfreie Prüfung (DGfZP) [The German Society for Non-destructive Testing]. *DGZfP-Fachausschuss für Zerstörungsfreie Prüfung im Bauwesen (AB)* [Technical Committee NDT in Civil Engineering], *Merkblatt für das Ultraschall-Impuls-Verfahren zur Zerstörungsfreien Prüfung mineralischer Baustoffe und Bauteile 1999, überarbeitete Auflage* [Guideline for the ultrasonic pulse method for non-destructive testing of mineral construction materials and elements] (Merkblatt B 4), 1999

Federal Highway Agency. Federal Lands Highway Program: Geophysical Methods. 2003 www.cflhd.gov/agm/engApplications/BridgeSystemSubstructure/212BoreholeNondestMethods.htm

Gaba AR, Simpson B, Powrie W & Beadman DR. *Embedded retaining walls: guidance for economic design.* CIRIA Report C580. London, CIRIA, 2003

German Society for Geotechniques (Working Group 2.1) Recommendations for static and dynamic pile tests. 1998. Available in German/English edition

Hanisch J, Katzenbach R & König G. *Kombinierte Pfahl-Plattengründungen.* Berlin: Ernst & Sohn. 2002

Holeyman AE. Keynote Lecture: Technology of pile dynamic testing. In: Barends FBJ (ed) *Proceedings 4th International Conference on the Application of Stress Wave Theory to Piles,* The Hague, The Netherlands, 21–24 September 1992. Rotterdam, Balkema, 1992. pp 195–215

ICE. *Specification for piling and embedded retaining walls.* London, Telford, 2007

Katzenbach R & Moormann Ch. Instrumentation and monitoring of combined piled rafts (CPRF): state-of-the-art report. *Proceedings. 6th International Symposium on Field Measurements in GeoMechanics (FMGM 2003),* 15–18 September 2003, Oslo, Norway

Katzenbach R, Bachmann G, Boled-Mekasha G & Ramm H. 2005. The Combined Pile Raft Foundations (CPRF): an appropriate solution for the foundation of high-rise buildings. *Proceedings Geotechnics in Urban Areas, 27–28 June 2005, Bratislava, Slovakia*

Kirsch F & Klingmüller O. *Erfahrungen aus 25 Jahren Pfahl-Integritätsprüfung in Deutschland* [25-years experience with pile-integrity testing in Germany] in Bautechnik 80. Heft 9. 2003. pp 640–650

Klar A, Frydman S & Baker R. Seismic analysis of infinite pile groups in liquefiable soil. *Soil Dynamics and Earthquake Engineering* 2004: **24:** 565–575

Komurka VE, Wagner AB & Edil TB. *Estimating of soil/pile set-up. Wisconsin highway research program, N° 0092-00-14.* 2003

Krause M, Mielentz F, Milmann B, Streicher D & Müller W. Ultrasonic imaging of concrete elements: state of the art using 2D synthetic aperture. In: DGZfP (Ed) *International Symposium Non-Destructive Testing in Civil Engineering (NDT-CE),* Berlin, Germany, 16–19 September 2003. Proceedings on BB 85-CD, V51, Berlin, 2003

Lee KL. Adhesion bonds in sands at high pressures. *ASCE Journal of Geotechnical Engineering* 1977: 103 (GT8): 908–913

Littlechild BD, Plumbridge GD & Free MW. Shaft grouted piles in sand and clay in Bangkok. *Proceedings 7th International Conference and Exhibition on Piling and Deep Foundations,* DFI, Vienna, 1998. pp 1.7.1–1.7.8

Long M. Database for retaining wall and ground movements due to deep excavations. *ASCE Journal of Geotechnical and Geoenvironmental Engineering* 2001: **127** (3): 203–224

Long JH, Kerrigan JA & Wysockey MH. Measured time effects for axial capacity of driven piling. *Transportation Research Record* 1663, Paper n° 99–1183. 1999. pp 8–15.

Middendorp P. Keynote lecture. Statnamic: the engineering of art. *Proceedings 6th International Conference on the Application of Stress-wave Theory to Piles.* São Paulo, Balkema, 2000

Ministère de l'Equipement, du Logement et des Transports. 'Fascicule 62'. *Règles Techniques de Conception et de Calcul des Fondations des Ouvrages de Génie Civil* (Technical Rules for the Design of Foundations of Civil Engineering Structures). Fascicule 62 - Titre V du Cahier des Clauses Techniques Générales, December 1993

Potts DM & Zdravkovic L. *Finite element analysis in geotechnical engineering - application.* London, Telford, 2001. 427pp

Poulos HG. Analysis of soil extraction for correcting uneven settlement of pile foundations. In: Leung CF et al (eds) *Proceedings 12th Asian Reg. Conference in Soil Mechanics Geotechnical Engineering,* Singapore, 2003. Vol 1, pp 653–656

Poulos HG. Pile behaviour: consequences of geological and construction imperfections. Report on 40th Karl Terzaghi Lecture. *Journal of Geotechnical and Geoenvironmental Engineering* 2004a: **131** (5): 538–561

Poulos HG. Control of settlement and load distribution in pile groups via stiffness inserts. *Proceedings Symposium on Recent Developments in Foundation Practice,* Center for Research and Professional Development, Hong Kong, 2004b

Powell JJM & Brown MJ. Statnamic pile testing for foundation reuse. In: Butcher AP, Powell JJM & Skinner H (eds). *Reuse of Foundations for Urban Sites: Proceedings of International Conference,* BRE, Watford, 19–20 October 2006. EP 73. Bracknell, IHS BRE Press, 2006. pp 223–236

Powell JJM & Skinner HD. Capacity changes of bored piles with time. In: Butcher AP, Powell JJM & Skinner H (eds). *Reuse of Foundations for Urban Sites: Proceedings of International Conference,* BRE, Watford, 19–20 October 2006. EP 73. Bracknell, IHS BRE Press, 2006. pp 237–248

Powell JJM, Butcher AP & Pellew A. Capacity of driven piles with time: implications for re-use. In: Vaniceck et al (eds) *Proceedings XIII ECSMGE,* Prague, 2003. Vol 2, pp 335–340

Rankka W & Holm G. Non-destructive methods for testing precast concrete piles under existing buildings. IIn: Butcher AP, Powell JJM & Skinner H (eds). *Reuse of Foundations for Urban Sites: Proceedings of International Conference,* BRE, Watford, 19–20 October 2006. EP 73. Bracknell, IHS BRE Press, 2006. pp 133–146

Rausche F, Goble GG & Likins GE. Dynamic determination of pile capacity. *ASCE Journal of Geotechnical Engineering* 1985: **111** (3): 367–383

Ravaschio P, Odasso A & Parker, EJ. *GOPAL Project Final Report.* MAS3-CT97-0119, Doc. N° 96-322-H13, European Commission, Brussels, Belgium. November 2000, 51 pp

Schickert M, Krause M & Müller W. Ultrasonic imaging of concrete elements using reconstruction by synthetic aperture focusing technique. *Journal of Materials in Civil Engineering* 2003: **May/June**

Schickert G, Krause M & Wiggenhauser H. ZfPBau: Kompendium

Skov R & Denver H. Time-dependency of bearing capacity of piles. *International Conference on Application of Stress-wave Theory to Piles,* Ottawa, May 1988

St John HD & Chow FC. Reusing piled foundations: two case studies. In: Butcher AP, Powell JJM & Skinner H (eds). *Reuse of Foundations for Urban Sites: Proceedings of International Conference,* BRE, Watford, 19–20 October 2006. EP 73. Bracknell, IHS BRE Press, 2006. pp 357–374

Taffe A, Krause M, Milmann B & Niederleithinger E. Assessment of foundation slabs with US-echo in the re-use process. In: *Proceedings International Conference on Concrete Repair, Rehabilitation and Retrofitting,* 21–23 November 2005, Cape Town, South Africa. Rotterdam, Balkema, 2005

Tester P D & Fernie R. A case study of total foundation reuse for a car park in Coventry, UK. In: Butcher AP, Powell JJM & Skinner H (eds). *Reuse of Foundations for Urban Sites: Proceedings of International Conference,* BRE, Watford, 19–20 October 2006. EP 73. Bracknell, IHS BRE Press, 2006. pp 375–384

Thompson CD & Thompson DE. Real and apparent relaxation of driven piles. *ASCE Journal of Geotechnical Engineering:* 1985: **111** (2): 225–237

Turner MJ. *Integrity testing in piling practice.* CIRIA Report 144. London, CIRIA, 1997

Vaziri M. Re-use of existing piles, Belgrave House, London. *Proceedings of ISSMGE, Osaka,* 2005, pp 2197–2204

Vaziri M & Windle J. Strategy for the reuse of existing piles: case study Belgrave House, London. In: Butcher AP, Powell JJM & Skinner H (eds). *Reuse of Foundations for Urban Sites: Proceedings of International Conference*, BRE, Watford, 19–20 October 2006b. EP 73. Bracknell, IHS BRE Press, 2006a. pp 47–58

Vaziri M & Windle J. Testing strategy for the reuse of existing piles: case study at Belgrave House, London. In: Butcher AP, Powell JJM & Skinner H (eds). *Reuse of Foundations for Urban Sites: Proceedings of International Conference*, BRE, Watford, 19–20 October 2006a. EP 73. Bracknell, IHS BRE Press, 2006b. pp 59–68

Verfahren der Zerstörungsfreien Prüfung im Bauwesen. Online database at BAM. http://www.bam.de/service/publikationen/zfp_kompendium/welcome.html

7 Design of new foundations for future reuse

7.1 Introduction

Future reuse of new foundations will reduce the need for building materials and other resources, and can help mitigate against sterilising sites for future redevelopment.

As identified in *Chapter 5*, before reusing a foundation it must be shown to have (or be designed to incorporate) sufficient capacity to carry the new load, durability for the new lifetime and a layout compatible with the new superstructure. Ways of providing this confirmation for a current design, and thereby improving future reuse potential include:

- use of instrumentation to monitor structural loading and performance of the foundations over their life-span,
- compiling and storing comprehensive records from the construction and operation of the building.

This chapter discusses how pre-planning, design, monitoring and documentation of a new foundation can be made in order to facilitate future reuse. The chapter includes some principles and state-of-the-art experience of monitoring acquired from the RuFUS project (Butcher et al 2006).

7.2 Design of new foundations for future reuse

7.2.1 Rethinking design working life

The best way to avoid filling the ground with deep foundations from each successive superstructure and stagnating urban sites is to extend the design working life of foundations. An extended design working life may allow for successive different superstructures, ranging from the addition of storeys to complete demolition and reconstruction.

The design working life is the assumed period for which a building is to be used without major repair (BS EN 1990). The reuse of foundations involves re-thinking design working life, either:

- by re-evaluation of the remaining design working life in the future, or
- by a choice of a longer design working life for the foundation than for the superstructure.

To allow for reuse of a foundation and an extended design working life after an addition of storeys, it might be enough to confirm that the foundation still has a suitable margin with the additional loading. Such a confirmation will be aided by any performance data from monitoring, detailed original design and as-built documentation.

7.2.2 Flexible foundation layouts

To extend the design working life of a foundation to allow for reuse after complete demolition and reconstruction of the superstructure, a foundation layout that allows different superstructure arrangements may be beneficial.

Ultimately, the design could be a 'platform' that can accommodate successive different superstructures. The use of such a platform would act to reduce significantly the construction programme for each new superstructure.

In practice, a designer may not be in a position to develop a platform solution, but layouts that provide some redundancy (such as pile groups or piled rafts) as well as consideration of basement walls that could be capable of carrying vertical loads in the future would be useful.

While it may be feasible to safeguard the foundations against a variety of future superstructures, generally it is not possible to anticipate all future building requirements and it is unlikely to be economical to design against all imaginable future requirements. Where there are planning restrictions, such as a maximum building height, the likely maximum structural loading may be more feasible to assess and safeguard against.

For structures such as bridges, however, ensuring that the foundations can accommodate alternative decks may help future reuse and provide significant advantage in necessarily limited transport corridors.

7.2.3 Other aspects of the design of new foundations for reuse

The design of a foundation for future reuse should take into consideration:

- access and markers for future testing of foundation elements,
- access for monitoring instruments, and
- design and construction informed by monitoring (observational method).

As a simple aid for future location and testing of foundations, markers could be used for each element. Examples of simple markers might be plates cast into the tops of piles. More sophisticated locators, such as passive radio tags, can also contain data on the foundation elements. Access for future testing (eg NDT) can also be considered during design. This may limit the amount of demolition required to reuse the foundations in the future.

Design can integrate access for installation and reading of monitoring instruments, although the construction process will require consideration.

Design and construction informed by monitoring can generate good records of foundation performance that can be used for reuse. This may be necessary to improve our understanding of current performance to extend into reuse, as well as direct confirmation of capacity for future reuse. If the behaviour of a building can be monitored in more detail than currently practiced, uncertainties in design parameters can be reduced during the construction stage and onwards and indeed may enable smaller design margins. This has been used successfully in the observational method, particularly where a number of temporary and permanent construction solutions such as retaining wall propping schemes are possible. Design and construction that uses monitoring may confirm the existence of extra capacity in the foundations that can be taken into account for later reuse purposes. This leads to the notion of 'Smart foundations' where use/reuse is tailored to particular loading conditions and foundation behaviour/response.

'Smart foundations', interrogated by strategically placed instrumentation, can record the changes in loading and corresponding settlement performance over the design working life of the foundations. When considering future foundation reuse, this monitoring information will provide additional confidence as to the condition and performance of the foundations and hence is likely to increase the level of reloading that can be assigned for an extended design working life. See *Section 7.3* for more details on monitoring.

Monitoring can be targeted at understanding loading/unloading conditions and load response and load distribution of a new or existing structure. Through monitoring, for example, confidence in an estimation of the distribution of load between slab and piles (Figure 7.1) may be increased. The increased confidence may allow for a higher future load than the original design load. Another example of how reserve capacity may be found is through the monitoring of the distribution and bearing of a load in a foundation wall (Figure 7.2). If the contact pressure between slab and soil is monitored and found to be higher than assumed during design, a reserve capacity exists. Also, if the deformations of the wall are monitored and calculated to correspond to a higher friction force between soil and wall, further reserve capacity exists. Provided that the structural capacity of the foundation elements is sufficient and no other failure mode is introduced, these reserve capacities may be used to carry future additional loads (see *Chapter 6)*.

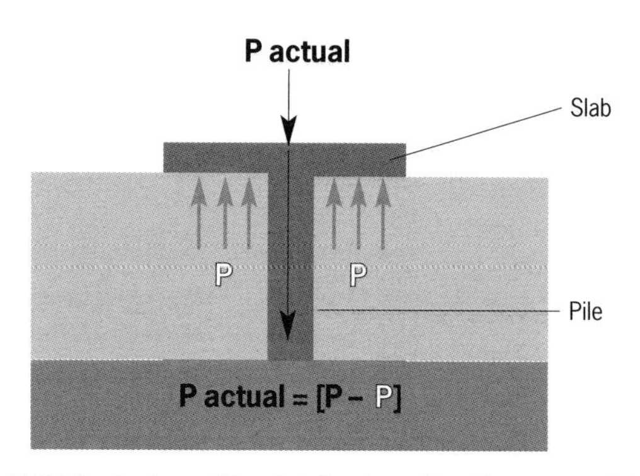

Figure 7.1 Monitoring of the distribution of load between slab and pile might allow for higher loads in the piles than the original design load

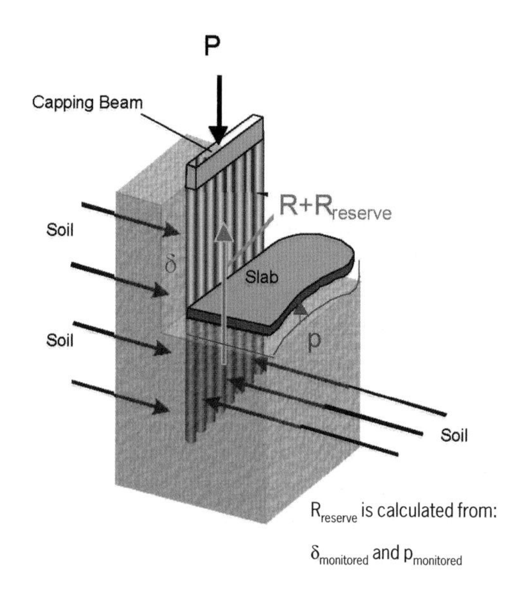

Figure 7.2 Monitoring of foundation wall deformation, δ, (indicating load/load transfer) might show that a reserve capacity exists

7.3 Monitoring of new foundations for future reuse

7.3.1 Who to monitor and why?

Monitoring new foundations provides information on foundation performance that can be used to improve the reuse potential for a site. Improvement in potential reuse will be of financial benefit to the client. It is important that the client buys into this and that the benefit is continuously assessed. Without the client's active support, monitoring proposals will not succeed. A major cost in monitoring is ensuring the robustness, survival and longevity of the instrumentation. For a successful outcome, the whole professional team must be involved (structural engineer, geotechnical engineer, main contractor, trade contractors, etc.). A 'Monitoring leader' recognised by all should be appointed and have sufficient support from the client hierarchy to drive the project through. Poor execution/interfacing of monitoring works will lead to unusable instrumentation. Generally, some allowance must be made for disruption in terms of time/programme and the associated costs.

7.3.2 What and when to monitor?

Instrumentation and monitoring systems included within the construction of new foundations may have the function of:

- verifying the settlement performance of the foundations as they are loaded,
- confirming that the loads applied to the foundations and the load distribution within the foundation system are within calculated limits,
- providing on-going measurements of the loading and settlement of foundations over the service life of the building.

For monitoring of structural elements, it is important to analyse the load transfer (and differential strain) between all elements of a foundation.

Ideally, multi-tasking sensors would report a variety of information. However, the need for robustness and economies dictate that these sensors are not yet available in civil engineering. Other industries are already developing multi-tasking sensors that report their condition and performance. It is anticipated that this type of system would ideally suit the monitoring of new foundations for future reuse.

All instrumentation layouts must be designed for redundancy. This is of particular importance in reuse as a number of construction and demolition phases may be involved. The information is of critical value to the reuse design so long-term readings must be cross-checked. Different parameters may require different instruments. It is important, however, to combine the instruments into a monitoring system, ie to correlate recorded data from different instruments and therefore consider redundancy.

Currently, strain is the main and easiest parameter to record. Strain is also indexed to most of the other parameters that are required (eg stress, load, bending moment).

Within a typical foundation for an urban building, the parameters shown in Figure 7.3 should be considered for monitoring.

The monitoring system requires recording/transmittal stations (RTS). RTS may be as simple as access points for manual readings, but may be logging facilities. These should be located where they can be accessed and protected during and after construction.

Monitoring will vary depending on the parameters to be measured and the requirements for the specific project. However, monitoring to enable reuse should ideally continue long term to provide data for the performance of the foundation over the life-cycle of the structure. At present, however, there is little long-term data for many of the monitoring parameters or instrument types and an outcome of the RuFUS programme has been to set up two sites (Bankside and Zlote) for investigation of these long-term issues. The sites are located in different ground conditions (over-consolidated clay at Bankside and cohesionless soil at Zlote). The investigation has already contributed to indexing of future sensor interpretation for these soil conditions.

As current sensors do not multi-task, integrity proofing (other than by inference) remains a separate task for future testing and may require careful consideration in the design of monitoring systems for reuse.

A key issue in monitoring is the requirement for a comparison between the readings taken and a hypothesis. This is required not only to design an appropriate scheme in the first place but also to assess the data as they are gathered. It is crucial to be able to make a detailed prognosis before accumulating a huge databank of measurements. For example, the client is intended to benefit from information collected in improving the potential reuse of his foundations. To ensure the data are useful, observations should be compared with the design forecasts and thresholds in any of the parameters (eg load/bending moment/movement). Design for reuse will use these comparisons to assess the design and performance of the foundation, any reserve capacities, any ability to relocate and transfer loading, requirements for control and assessment of element integrity.

In addition to team commitment to data acquisition, long-term commitment to data recovery is required. Progress is rapidly being made to allow wireless data transfer that can aid this process and facilitate access to the data by interested parties. Longevity of not only the instruments themselves but

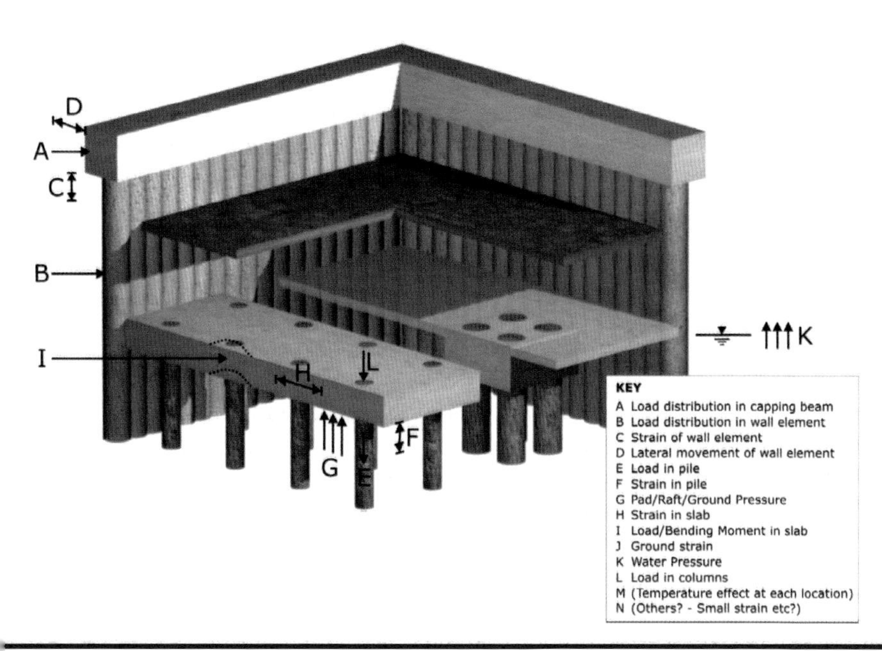

Figure 7.3 Example of parameters to be considered for monitoring

the data type and the databank collected (eg any requirements for electronics, etc.) should be considered.

7.3.3 How to monitor?

Practical aspects of monitoring

The instrumentation installation should be carried out by an experienced specialist in the appropriate sensor technology. Important issues for the monitoring leader are often the understanding and managing of the interfaces, including:

- connections from the monitoring element to the external wiring or monitoring interface where non-wireless sensors are used (eg from a VW gauge to the top of the pile there is potential for damage by pile breaking),
- link into RTS and physical access to RTS,
- remote access from RTS to interpreter's destination (eg transmittal to client/engineer, etc.)

The vulnerability of any system is highest in the temporary condition when a number of trade contractors are working in close proximity to the elements of the instrumentation.

Requirements of instrumentation systems

Experience within the RuFUS project and from other situations has shown that there are a number of key issues/potential problems with instrumentation.

The basic requirements of any sensor or monitoring system can be summarised as in Table 7.1 over the lifetime of a project. Based on the RuFUS project, the relative importance of each requirement during different stages is also given in Table 7.1. For example, consideration is taken of the expected time for use of a sensor. Short-term measurements are used for evaluation of real loads and deformations during the construction and only one or two years afterwards. Long-term measurements are used for evaluation of additional changes of load and deformations that may occur after the short term. In addition to the requirements of Table 7.1, it is also important to assign responsibility for the systems and the monitoring in the long term (eg to decommissioning of the building).

Remote interrogation, ie interrogation off the site, in the construction stage is valuable because regular monitoring and data transfer are easier, enabling quick action to be taken should some value (of settlement, load or other parameter) exceed a limiting value.

Sensors for monitoring instruments

To measure the parameters shown in Figure 7.3, one or several types of sensors may be used. It is important to consider the type of a sensor because different types fulfil the requirements of Table 7.1 to different degrees. The following main types of sensors are considered appropriate for reuse purposes:

- mechanical,
- magnetic,
- electrical,
- optical.

Monitoring instruments

The sensors can be combined into simple or complex instruments (eg fibreoptic cables combined into an inclinometer). The following main types of instruments are considered appropriate for reuse purposes.

- For levels and positions:
  - geodetic instruments,
  - fluid level meters, and
  - extensometers (of various types, examples of which are electric, vibrating wire, fibreoptic and magnetic),
- For inclinations:
  - inclinometers (of various types with various sensors; examples of types are: manual into borehole, automatic string into borehole, long electrolevel beams and local tilt meters; examples of sensors are: electromagnetic, vibrating wire and accelerometers),

Table 7.1 Importance of requirements for different stages

Requirement	Construction stage	Short-term stage	Long-term stage
Robustness	Very important	Very important	Very important
High resolution	Not so important	Important	Very important
Durability	Not so important	Important	Very important
Accuracy	Important	Important	Important
Economy (equipment, installation, monitoring)	Important	Important	Important
Remote interrogation possibility	Important	Important	Not so important
Calibration possibility	Important	Important	Very important
Compatibility with future electronic equipment	—	Important	Very important
Easy to use	Important	Important	Important
Measurements and documentation of activities in the surrounding environment	Important	Important	Important
Revision after evaluation of short-term measurement	—	—	Important
Redundancy (type, number)	Important	Important	Very important

Note: The importance is classified in a relative scale: not so important, important or very important

- For strains:
  - strain gauges (of various types, examples of which are electrical resistance, vibrating wire and fibreoptic),
- For pressures and force:
  - load cells (which can be based on strain gauges or pressure cells),
- For cracks:
  - width meters (of various types, examples of which are mechanical, electric, vibrating wire and fibreoptic).

Indirect measures are often evaluated from direct measurements (eg fibreoptic measurement of the distribution of strain along a pile can be used to show the presence of cracks.

Data handling within monitoring

The data handling within monitoring includes acquisition of data from the instruments, local storing (logging) of data and transmittal of data to a database. Many sensors can be read either by automatic or manual acquisition systems. Hence, in practice, the choice between automatic and manual system generally depends on the required frequency of readings. For readings once a week, month or year, automatic systems are often too expensive to install and maintain, and manual readings are carried out. For shorter reading intervals, from seconds to every few days, manual readings become too expensive and automatic systems are put in place.

Automatic acquisition systems of all sorts generally include a 'datalogger', with a link to a computer and database.

- Dataloggers are devices specially designed and built for the task of acquiring data. Some are more laboratory-oriented, and some are designed for site use. The datalogger should be robust as appropriate for the monitoring design. Modern loggers can be powered easily by solar panels.
- Link to computers: can be manual (the user comes and connects a computer to the datalogger to retrieve data, or automatic (link by cable, by radio, by phone, etc. to the computer).
- The computers used are generally standard, sometimes industrial versions for greater robustness.
- The database can be anything from a simple ASCII file or spreadsheet to store the data, to a fully dedicated and multi-options database and presentation system. Database 'duties' include calculating engineering values (derived values) from raw data, data storage, and, depending on the level of system complexity, data presentation to help the analysis, trigger alarms, provide data transmission to internet, reporting, etc.

Manual acquisition systems are generally based on a readout, whether electric or manual. The user takes readings that are either stored in the readout memory or hand written by the user. The results are either direct engineering values or raw values or spectrums that require further analysis or calculations. Finally, data are imported into a database.

Systems for data-handling within monitoring are often called recording/transmittal stations (RTS).

Examples of the application of monitoring instrumentation are given in *Appendix E*, Soletanche Bachy (1998), Dore et al (2002) and Katzenbach (2004).

Descriptions of the current technology for sensors and instruments for monitoring may be found in handbooks [eg Dunniclif (1994) or a DTI/CIRIA report (Buenfeld, to be published, 2007)]. The description given below is a summary of the experience from the RuFUS project concerning application of current technology and development of applications with new fibreoptics technology.

7.3.4 An evaluation of state-of-the-art monitoring instruments

The state-of-the-art sensors for 'smart foundations' are currently fibreoptic sensors. The following are some of the potential advantages of fibreoptic sensors over conventional sensors:

- immunity to electromagnetic interference,
- small size,
- light in weight,
- measurement of different parameters (strain, temperature, vibration and specified chemicals),
- multiplexing capability enabling distributed mapping of a structure.

Fibreoptic sensors need adequate protection systems to ensure robustness and longevity of readings. While fibreoptic sensors are currently still in a research and prototype phase, the RuFUS conference proceedings (Butcher et al 2006) give a number of examples of their application in construction.

Because fibreoptic sensors are still not common in foundation engineering, short descriptions of the main fibreoptics systems for monitoring are given in this *Handbook*. In the sections below, descriptions are given of the two systems used in the RuFUS project, while in *Appendix E* the other main systems are described. The descriptions are based on López-Higuera (2002). The RuFUS project has enabled development and validation of the Bragg grating and Brillouin fibreoptic systems (see *Appendix E*) and this gives a high level of confidence in their performance. The systems differ in that Fibre Bragg gratings are point sensors, whereas Brillouin sensors give a continuous trace averaged over discrete lengths. Both systems are rapidly developing. The advantages and disadvantages as identified by the RuFUS project may be found in Table 7.2.

Table 7.2 Comparison between fibre Bragg grating sensors and Brillouin sensors

Brillouin	Bragg grating
Low-cost installed sensor (cable is the sensor)	Higher costs for Bragg grating due to licensing
5–20 min period to measure strain	'Real time' readings
Expensive (~£50 K) analyser	Cheap analyser

For monitoring of new foundations for reuse, sensor longevity is a paramount requirement. Fibreoptic cables have demonstrated longevity in harsh environments and in theory this should be transferable to civil engineering applications. To date there have been no adverse reports of sensor drift and deterioration over the past 5–10 years. This compares favourably with traditional transducers and other strain measurement sensors.

Fibre Bragg grating strain and temperature sensor
A Fibre Bragg grating sensor is a single-mode telecommunications optical fibre with a region (or grating) of alternating bands of high and low refractive index. The region is created by exposure of the original fibre to UV laser light. The reflection characteristics of the region vary with strain and temperature. When the region is exposed to light, it is possible to measure strain and temperature through analysis of the spectrum of the reflected light. The spectral analysis can be done by a Fabry-Perot cavity or a spectrometer. To correct strain values from the effects of temperature a free (from strain) reference sensor is needed. Many sensors can be created along the same fibre and tuned to reflect at different wavelengths. The different tuning allows for identification of different sensors. The number of sensors along a fibre is typically 4–16.

For embedded sensors, the following have to be quantified before interpretation of results:

- bond strength,
- long-term stability of bond,
- varying bond strength along a sensor (for a number of gratings),
- irregularities of surrounding matrix (eg leading to transverse point loads),
- the limit of strain deformation.

The limiting number of sensors (locations) is typically 40 for current technology analysers.

Brillouin distributed temperature and strain sensor
Acoustic waves (phonons) produce a periodic modulation of the refractive index of a fibre. A so-called Brillouin scattering occurs when light propagating in the fibre is diffracted backwards by the fibre. This process is called spontaneous Brillouin scattering. The main challenge is the extremely low level of the detected signal (which leads to sophisticated signal processing) and long integration times.

Measurement of distributed strains requires a specially designed sensor and a mechanical coupling to the object to be measured along the whole length of the fibre.

Systems able to measure strain or temperature variations of fibres with length up to 50 km and spacing in the metre range have shown their potential in field trials. For strain measurements over this range it has practically no rival. Commercial systems based on spontaneous Brillouin scattering are available.

For Brillouin optical time-resolved spectroscopy (BOTDR), the analyser is expensive but standard fibre can be used. The fibre has to be pre-tensioned in order to measure compression. There are some questions of how to calibrate for temperature variation. Also, there is a need to understand fully the robustness and longevity of the method. The method seems favourable. Klar et al (2006) have stressed the advantages of a continuous monitored length compared with only local information when evaluating the strain along a pile.

Examples of the application of fibreoptic instrumentation in civil engineering applications are given in *Appendix E* and in papers presented at the *RuFUS* conference (Butcher et al 2006).

7.3.5 An evaluation of monitoring instruments used during the RuFUS project

Examples of instruments and how they meet the requirements of Table 7.1 according to the RuFUS experience are given in Table 7.3.

RuFUS experience with RTS systems (linked to the internet at Zloty, long-term data logging in Frankfurt) shows that RTS acquisition units may be expected to function at least within the short-term stage. A possible limitation to be considered is, however, the risk that future computers and computer codes may turn out to lack compatibility with existing computers and computer codes.

7.4 Documentation of new foundations for future reuse

Current recording, storage and longevity of design and construction data are sketchy. BS EN 1997-1 provides a more detailed framework of records to be kept with regards to the ground condition, design and as-built construction. For piles, BS EN 1997-1 specifies that the:

'installation of all piles is monitored and records are made as the piles are installed'. [Clause 7.9 (3)]

'Records should be kept for at least a period of five years after completion of the works. As-built records should be compiled after completion of the piling and kept with the construction documents'. [Clause 7.9(5)]

With regards to construction records,

'Inspection records during construction are required'. [Clause 4.2.2(5) & (6)]

'More important documents should be stored for the lifetime of the relevant structure'. [Clause 4.2.2(8)]

While these requirements go some way to specifying what records should be kept, further clarity on the specific records, responsibility for storage and storage lifetime need further clarification to safeguard new foundations for potential future reuse.

Typically, design information has been kept at the discretion of the firm of consulting engineers. Likewise, construction records are usually only saved for a finite time

Table 7.3 Evaluation of instruments

Parameter	Instrument	Sensor components	Construction element (example)	Maximum stage	Reading
Position	Geodetic	Optical or laser	Surface of foundation	Long term	Manual or remote
	Extensometer	Mechanical	Pile	Long term	Manual
	Extensometer	Magnetic and electrical (magneto-resistive)	Soil layer	Short term	Manual
Inclination	Inclinometer	Magnetic and electrical (magneto-resistive)	Diaphragm wall	Short term	Remote
	Inclinometer (strain), type Smartrod	Optical fibre (Bragg gratings)	Diaphragm wall	Short term	Manual
Distance	Strain gauge	Mechanical and electrical (vibrating wire)	Raft (rebar), pile, secant wall, capping beam	Short term	Remote
	Strain gauge	Optical fibre (Bragg gratings)	Pile	Short term	Manual
	Strain gauge	Optical fibre (Brillouin)	Pile	Short term	Manual
Pressure	Load cell	Electric (strain)	Pile head, head of ground anchor, end of strut, cap–soil interface	Short term	Remote
	Load cell	Hydraulic and electrical (electromagnetic)	Raft–soil interface	Long term	Remote
	Load cell	Hydraulic and electrical	Pile head, pile toe	Long term	Remote
	Load cell	Hydraulic and pneumatic	Pile head, pile toe	Long term	Remote
	Pressure cell	Hydraulic and electrical (piezometer)	Raft–soil interface	Long term	Remote

by contractors. Space is at a premium and old records are often sorted through and disposed of once the decision has been taken that they are no longer required.

When considering foundation reuse, information on the existing foundations will be very valuable to the design team. However, there are currently no requirements or guidance on the preservation of records of foundations to aid future reuse. During pile construction, records are submitted by the main contractor and the design engineer. This usually incorporates the as-built information on a standard form to ensure that construction is complying with design. If kept, this information is a valuable resource for the new design engineer when considering foundation reuse.

As-built information is a much more reliable indicator for what old foundations remain in the ground than the original pre-construction design calculations and drawings. Using the as-built information, any deviation from design or defects in the piles can be evaluated for suitability for the new scheme.

To maximise the reuse potential and the resale value of a site, property owners should ensure that good records are collected and collated for their buildings so that future developments on those sites can benefit from the information. Once compiled, these records should be stored by the property owners in a secure location so that they will not be lost as a result of events that may happen in the future such as the designer or contractor going out of business. *Property owners should consider that these documents are as important as details of the building insurance, property deeds, etc.*

7.4.1 Requirements on documentation

When starting the RuFUS project it was discovered that one of the main barriers to reuse of foundations was the lack of detailed information about foundations and verification of any existing records. To facilitate reuse better in the future, some ideas on the constituents of an information system for foundations are presented in this chapter.

To enable wider reuse of foundations in the future some information from the design and construction of foundations is vitally important. Key information may include, for example, bearing capacity of old piles, whether the pile is single or in a group, if the pile is loaded transversally or not or whether the pile is loaded in compression or tension. The range of information that is valuable for reuse of a foundation includes the parameters identified in Box 7.1. This includes as-built as well as design information.

It is recommended that for sites where archaeology has been preserved beneath building foundations, the documentation should be placed with the appropriate local historical record centre (eg with the sites and monuments record in the UK). This will ensure that this information is available to archaeologists trying to preserve this archaeolgical material, even if the new owner of a property is not interested in keeping the data.

Much information is specific to each country and therefore professionals operating in each country should recognise what is important. For example, in Italy and Greece it is necessary to consider earthquakes when designing a foundation, while this is not necessary in the Scandinavian countries. On the other hand, in, for example, Sweden, it is

Box 7.1 Valuable information on new piles that should be stored

Program stage
Geological information
Geotechnical information
Groundwater level
Groundwater quality
Contaminated soil
Site conditions

Design stage
Design philosophy
Design codes
Design calculations
Necessary bearing capacity
Force combinations applied on each pile
Pile data
Settlement limitations
Required testing and monitoring
Protocol for foundation records

Construction stage
As-built documents
Non-conformance reports
Construction documents
Programme of piling works
Plant and equipment
Details of pile integrity testing
Results of pile load testing
Working documents
Site records
Pile installation records
Effects on nearby foundations and structures
Results from monitoring

Control stage
As-built drawings
Maintenance records
Environmental changes
Inspections
Pile behaviour
Service life measurements
Structural alterations

necessary to consider extreme cold, causing deep frost, so information relevant to this is important.

7.4.2 Information system for new foundations

The aim of the RuFUS work has been to build the general structure of an information system, and not a direct application. The approach has been to use a process and product model structure. To see what data need to be handled in an information system for a foundation, so that it enables reuse of the foundation in the future, it is important to outline the foundation process (Figure 7.4). An example of typical data derived from current practice is given in *Appendix F*.

At each stage of the foundation process several documents and information are delivered and stored. These documents and information are often used within the next stage of the process (eg from the design stage, design calculations, reports, drawings, etc. are delivered to the construction stage). This information is valuable for the future reuse of the foundation. When reusing a building and foundation, the design and construction process starts again, from the program stage.

'Governing information' is information that is essential for the whole building, not only the foundation (eg city plans, regulations, etc.). 'Supporting information' is information that contributes to the actual stage of the process.

The different parts of a foundation can be divided into structural elements and spaces, according to IFC-Industry Foundation Classes. It is a framework of classes that can be applied to objects of interest for the construction and real estate industry and provides a foundation for a shared project model. Structural elements can be described as part of the building that alone or together with other structural elements are required to provide strength and stability. Space is 3D and is built up by structural elements. The foundation can be divided to fit into an information structure; the Swedish system (Figure 7.5) exemplifies this model.

Each structural element or space element can be further divided and arranged into sub-groups. An example of how pile information can be arranged using a building resource model and the national Swedish classification system is shown in Figure 7.6. For each structural element properties are specified and the value and unit are also specified. Only some properties are shown in this example.

Each element in the structure will receive certain attributes that further describe the element. Each pile is given a unique identification number when the design is started (or the project begins) and this number follows the pile through the whole process, from design to construction, to service life and reuse.

The documents connected to the foundation process are best managed by using an electronic document management system (EDMS) and metadata. Metadata is a description of a document and its content. The document together with its contents are stored in its original place and its address is

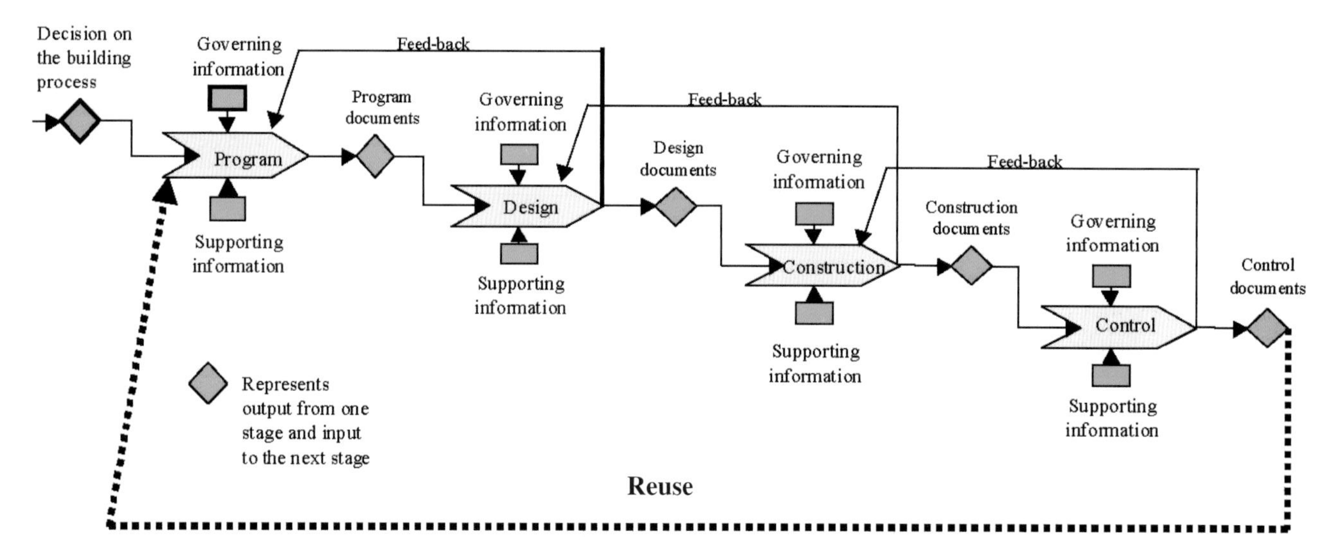

Figure 7.4 The foundation process

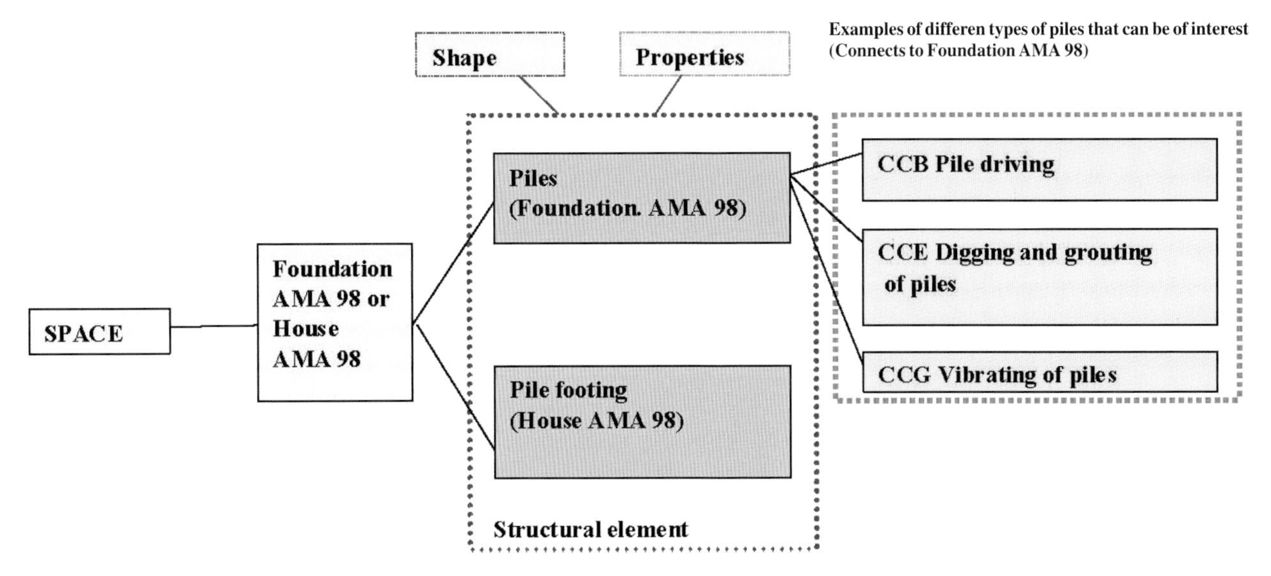

Figure 7.5 Example of how a pile can be specified by properties (Swedish system)

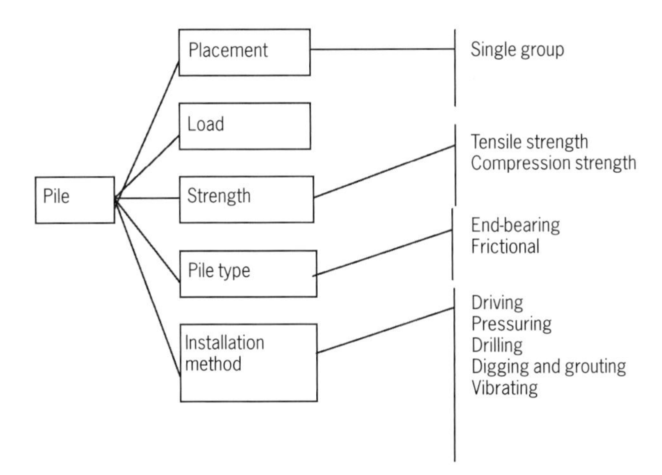

Figure 7.6 Example of how a pile can be specified by properties

registered in the metadata database (a kind of address register, rather than an actual database). Metadata is the best way to enable access of documents in databases and information systems and since it uses the extensible file transmission system of XML (eXtensible Markup Language), it is easy to expand the system. XML is being used for transmission and exchange of database information. It has many advantages compared with HTML and is suitable for information exchange between different types of computer system and to structure documents and information.

Collection of data that is important for reuse of the building and the foundation is best made by the property owner (the client). It is important that the property owner ensures that technical data along with other building records (title deeds, sales information, insurance, etc.) are collected and collated for their buildings so that future reuse of the site can benefit from the information. Paper records should always be kept. However, electronic records are also recommended. By using this structure, all information follows the property when selling or buying it.

It is important that an information system is kept simple, so that it is easy and efficient to use.

7.4.3 Subsystems

There are several subsystems already in existence that influence an information system. These may include national specification and classification systems, BS EN 1997-1, the execution codes, QA-systems, national documentation systems, the building project, etc. (Figure 7.7). These systems are clearly developed and are already being used. It is therefore essential that an information system for foundations is connected to these systems in one way or another.

Governing information from, for example, the Eurocode affects the way the information is handled in the information system. National specification and classification systems (eg the Netherlands STABU system, the British NBS system and the Swedish BSAB 96 system) give guidance on how to structure the information from the building process. Each company has quality assurance (QA) systems that are a major part of the information process and it is important that the QA system is linked to the information system.

FI 2002 (Facilities Management Information 2002), a Swedish research and development project aims to develop suitable information structures for the facilities management area. This project developed a number of useful outputs, eg process analyses and models describing the whole scope of facilities management (FM), both management and core processes. To secure a unitary terminology in the sector a coherent vocabulary (iea list of terms with definitions) was

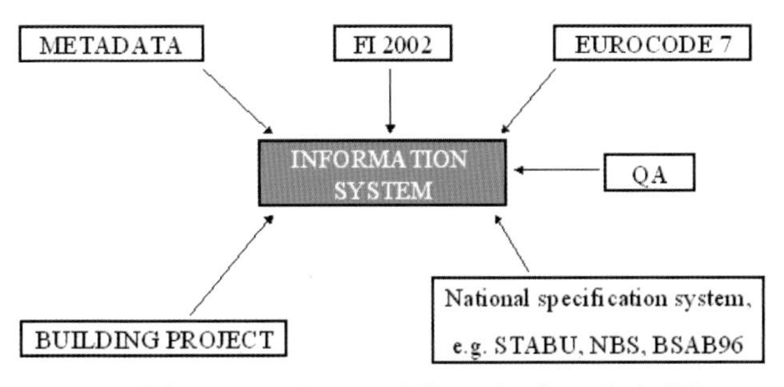

Figure 7.7 Connection between information from the building project and existing systems that needs to be considered in the foundation information system

developed. A set of object models (conceptual models), all of which together constitute the structure of a common database suitable for handling shared FM information (a data repository). This set of coherent object models is referred to in the project as the 'FI 2002 model'. By applying this object model to the foundation process, this information could be managed more easily, together with other important information for the building.

The information system handles information from the specific building project and its metadata. Metadata is used for catalogue information (eg electronic documents) and can be described as information about the information. Metadata is necessary to be able to access documents in databases and information systems and through this path to make them available even if they are part of large library of documents.

There exist international standards for handling and indexing information and transferring documents, and also for handling metadata. It is important that existing standards are used when building an information system that will enable reuse of foundations in the future.

7.5 Key points

- Future reuse of new foundations limits the requirements of raw material and thereby contributes to a sustainable society.
- Design of foundations that allows for flexibility in the layout of the structure that it can carry, increases the potential for future reuse of new foundations.
- Monitoring reduces the uncertainties in the distribution of loads and may identify reserve capacities in a foundation. This can be taken into account when reuse of a new foundation is considered in the future.
- Fibreoptic sensors show promise for monitoring of new foundations for future reuse.
- Systematic collection and storage of information from the foundation works are vitally important in enabling future reuse of any new foundations.
- In order not to lose any information, paper documents should be kept, although additionally the incorporation of information into an electronic system has advantages.
- Property owners are advised to consider the documentation from the foundation works as being as important as details of the building insurance, property deeds, etc.

7.6 References

British Standards Institution. BS EN 1990: 2002 *Eurocode. Basis of structural design*

British Standards Institution. BS EN 1997-1: 2004 *Eurocode 7. Geotechnical design. General rules*

Buenfeld N et al. *Intelligent monitoring of the deterioration of concrete structures.* DTI/CIRIA Report. London, CIRIA. To be published 2007

Butcher AP, Powell JJM & Skinner HD (eds). *Reuse of Foundations for Urban Sites: Proceedings of International Conference*, BRE, Watford, 19–20 October 2006. Bracknell, IHS BRE Press, 2006

Dore V, Amicel Y & Salvi G. *Le grand palais travaux de rénovation des fondations et instrumentation de côntrole.* Travaux no 791, novembre 2002

Dunnicliff J. *Geotechnical instrumentation for monitoring field performance.* Chichester, John Wiley, 1994

Habel WR. Long-term monitoring of 4,500 kN rock anchors in the Eder gravity dam using fibre-optic sensors. Article presented by Federal Insitute for Materials Research and Testing (BAM), Berlin, Germany, 2000 or later

Katzenbach R. *Foundation instrumentation: case studies.* Report no. T75-0221/G63.2 from Faculty of Civil Engineering and Geodesy at Technische Universität, Darmstadt. RuFUS project, 2004

Klar A, Bennett P & Soga K. The importance of distributed strain measurement for pile foundations. *Proceedings of ICE, Geotechnical Engineering* 2006: Special edition: Innovations in deep foundation: design and construction

López-Higuera JM. *Handbook of optical fibre sensing technology.* New York, Wiley, 2002

Appendix A: Case histories

1 Juxon House, London, UK

The previous 1960s building was constructed on piles with up to 3.2 m diameter under-reams. Construction records showed abandoned and remedial piles and included detailed logging of ground conditions.

The site was redeveloped on a slightly larger footprint with larger floor spans. The concentration of piles and under-reams made it difficult to squeeze in additional piles. It would have been impossible to found the new building on a completely new set of pile foundations without expensive, high-risk pile extraction and potential disturbance of archaeological remains.

The preferred foundation scheme was the reuse of the piles with some new ones where building loads were greater than before or where there were no existing piles. A new ground investigation was carried out. Pile condition was assessed by exposing some of the pile heads and taking and testing concrete and groundwater samples. Petrographic testing and analysis by concrete specialists provided assessments of concrete condition and durability. The reused pile heads were exposed during construction and Further concrete testing and integrity testing was carried out. Large pile caps redistribute loads between the new and existing piles so an iterative soil–structure interaction analysis was conducted on the composite pile groups to check the predicted loads, settlements and bending moments in the raft.

See also **St John HD & Chow FC.** Reusing piled foundations: two case studies. In: Butcher et al (eds) *Reuse of Foundations for Urban Sites: Proceedings of International Conference*, 2006. pp 357–374

2 Thames Court, London, UK

The 1960s building was founded on a thick concrete raft on straight-shafted bored piles through alluvial deposits and Thames Gravel into the London Clay. The piles were installed through the raft.

When the site was redeveloped in the 1980s, the new building plot included a substantial area outside the raft. A scheme with a shallower basement was eventually adopted which used the existing raft and constructed a new basement next to it linking to the raft. Additional piles, longer than the original ones, were added to the raft at isolated points beneath heavily loaded columns. The raft was sufficiently stiff for it to behave more or less rigidly. Therefore, the new piles had to mobilise a large enough load at low deflections to provide sufficient support to ensure that the steel in the raft was sufficient. The detailing of the link between the raft and the new pile was a critical issue.

3 Empress State Building, London, UK

The 30-storey Empress State Building, built in the 1960s, was founded on a basement raft supported by under-reamed piles. The extensive refurbishment programme included a 5.5 m wide horizontal extension over 27 floors and a three-storey vertical extension of the existing building. Rather than knocking the old building down and erecting a new one, the life of the building has been extended.To limit the increased loads on the existing piles new cantilever ground beams on new stiff piles were built.

See also **St John HD & Chow FC.** Reusing piled foundations: two case studies. In: Butcher et al (eds) *Reuse of Foundations for Urban Sites: Proceedings of International Conference*, 2006. pp 357–374

4 Holborn Place, London, UK

In the course of the redevelopment of the previous multistorey office building over a 4-level basement, the superstructure was demolished and rebuilt. New basement floors were built within the existing retaining walls and bottom basement raft.

Detailed investigations of the ground and groundwater conditions were undertaken, along with investigations of the existing pressure relief system and assessments of the existing structure. Hand calculations and finite element (FE) analyses were used to help understand how the current basement walls were functioning and what support they would need both during demolition and reconstruction, and in the long term. The walls had been designed for much lower earth pressures than would be considered appropriate now. Monitoring was carried out to check the wall deflections during demolition and reconstruction.

The successful reuse of the retaining walls resulted in significant time, cost and material saving and reduced the need for road closures around the site.

5 Tower Place, London, UK

Bowring Tower, a 14-storey building with up to 4 basement levels built in 1960, has been replaced by the Tower Place office development. This new development consists of two 6-storey buildings linked by an atrium with up to 3 basement levels.

Drawings of the existing bored cast-in-place concrete piles in London Clay were available. Based on pile tests on newly constructed piles, the expected axial capacity of the old piles was re-calculated and the pile design methodology was revised. As only limited information on the reinforcement of the existing piles was available, the lateral loads on the new building are not carried by these piles.

To assess the old piles, the exposed surface of the concrete was examined visually. Vertical coring was performed to assess the consistency and strength of the concrete, and to identify discontinuities, compaction quality and other features of concern (eg chemical attack). Concrete compressive strength and chloride content tests were carried out. Any unusual features, current deterioration or potential deterioration of concrete were assessed by means of petrographic examination. The exposed surface of reinforcement was examined visually. The tensile strength of the reinforcement was tested and the length of reinforcement determined by cover meter with ultra-sensitive antenna.

Other important issues during the construction works were heave and the prevention of damage to the piles caused by construction, operation, demolition of the old building and the construction and operation of the new building.

6 Tobacco Dock, London, UK

The former wharf, Tobacco Dock, has been refurbished as a 2-level shopping mall. The old foundations are brick piers on timber grillages supported on timber piles.

To assess the axial pile capacity 8 load pile tests to failure were carried out. Consistent behaviour up to about 100 kN was observed which is the approximate load of the self weight of the warehouse. After the tests, the piles were pulled out to assess the condition of the timber visually and to determine the material properties, such as the strength of the timber, growth ring, slope of grain, density, specific gravity and moisture content.

7 Arup Fitzrovia, London, UK

Six properties have been refurbished. All but one of these buildings are founded on piles. The pile loads increase by generally 14–19% with 2 pile groups increased by up to 22–31%. No structural problem during the lifetime of the existing building occurred. Additional information about the old piles was scarce except for the as-built piling layout of two buildings, pile diameters and two historical pile test data. Detailed information on pile reinforcement was not available.

It was not standard practice in the 1950s and 1960s to reinforce bored piles over the full length; usually, it was the upper portion of the pile that was reinforced. The reinforcements were not thought to be placed to deal with heave. Assessment of potential excess capacity of piles was carried out based on the information obtained from new ground investigations. Previous pile tests were reviewed and compared with old design lines.

Current design methodology was used in the assessment of the capacities of the old piles. In relation to the old pile load, a factor of safety of over 2 was calculated. On three piles visual examination of the exposed surface of the concrete was performed. Seven cores were taken from these three piles for visual examination, petrographic analysis and strength testing. Nine further integrity tests were conducted to obtain information on the pile toe level, any indication of reduction in pile cross-sectional area and defects in piles such as cracking.

See also **Anderson S, Chapman T & Fleming J.** Case history: the redevelopment of 13 Fitzroy Street, London. In: Butcher et al (eds) *Reuse of Foundations for Urban Sites: Proceedings of International Conference*, 2006. pp 311–320

8 The Law Courts, Marseille, France

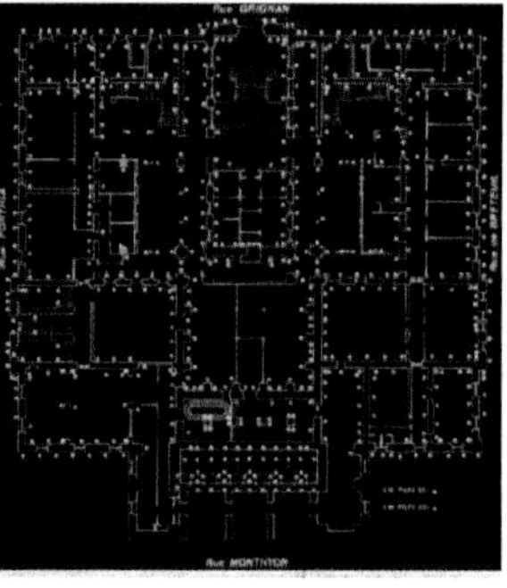

The Marseille Law Courts' building is more than a hundred years old. It has 3 to 4 levels and measures about 60 x 50 m. Loads are carried by load-bearing walls and columns resting on large concrete footings 2–6 m high, founded on wooden piles driven into sandy-clayey silt. A drop in the groundwater table had caused deterioration and even destruction of the wooden piles, and settlement of the silt.

In 1974, a decision was made to give the building stable foundations without interrupting the activities inside. The solution adopted was to transfer loading to a formation of compact marl at a depth of about 11 m by using IM piles (I = injection, M = metal). Each of these metal piles, with a capacity of 32 or 50 tons, was pressure grouted and sealed at the base by injection into the compact marl to a maximum length of 10 m. The piles of the bearing walls were slightly inclined and placed in a staggered pattern on either side of each wall. The IM piles consist of two rods and a sleeve pipe allowing pressure grouting for anchorage. The steel is stressed to half its elastic limit. The rods are protected by two coats of anti-corrosion paint. The elements, each 2.5 m long, are connected to one another by sleeves and welds. The piles are attached to the masonry by injecting grout.

9 Centre Tertiaire, Lille, France

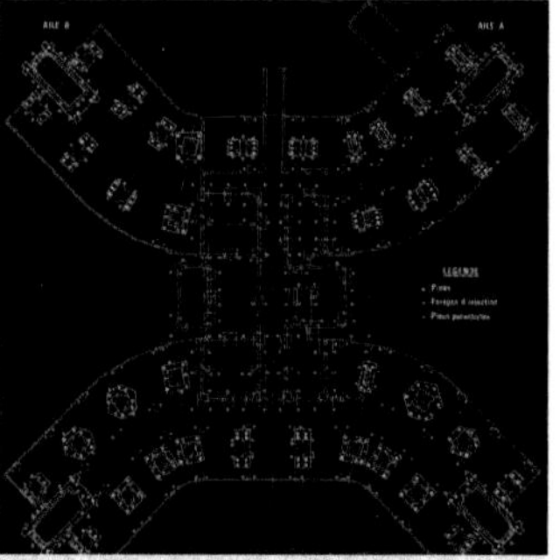

This 14-storey office building which was built on piles suffered from serious differential settlements. After a soil investigation campaign, the decision was to underpin the entire building by means of 335 IM piles (I = injection, M = metal). These friction piles with a length of 22–35 m were connected to the existing structure of the pile caps by additional prestressed concrete components. To protect the building from further damage, IM piles were used as an immediate measure.

10 Grand Palais, Paris, France

Built on a site near the River Seine, Paris' Grand Palais was constructed on a foundation of 3300 oak piles, a common practice for waterlogged sites when construction began in 1897 for the 1900 exhibition.

Flow management works on the Seine led to a drop in groundwater levels. This inevitably affected the integrity of the piles, causing differential settlement of between 100 mm and 120 mm. The mixed nature of the ground had allowed the original designers of the Grand Palais to use piles only under the south side facing the Seine, where they compensate for the instability of the alluvial deposits.

Two foundation techniques will be used in the restoration because of the variable nature of the site. In the piled area a diaphragm wall is being built. This will provide a base for the load-bearing steel supports as well as freeing up the interior of the basement to provide a space for possible future development. On the northern side, the ground has better load-bearing qualities and better homogeneity. Jet grouting will be used here to form 2100 columns of between 1 m and 1.40 m diameter to underpin the foundations. The columns are formed at an angle to pass under the existing structures.

The building's location in the heart of Paris, plus its complex layout, have provided further challenges for the project team. The museum galleries were open to the public throughout the works, restricting access to some parts of the site, and requiring noise levels to be kept to 50 dB(A) in the vicinity of the galleries.

11 Helgeandsholmen, Stockholm, Sweden

The Parliament Buildings in Stockholm were built in 1895–1905 on the island of Helgeandsholmen. The heavy Parliament buildings were founded on about 15000 driven timber piles.

During repair works in 2002, two of the existing 100-year-old timber piles were pulled up and their durability was investigated. Occasionally, the water level is just around the level of the mattresses and the pile heads. Due to land heave after the land glacial period (4 mm per year in Stockholm) there is a need to protect the mattresses and the timber piles from coming above the water level in the future.

Several types of investigation were performed on the two timber piles which showed that the core wood had not been attacked by bacteria or fungus after 100 years. However, the splint wood was attacked, most severely close to the pile surface and decreasing severity towards the core wood. The attack, principally by bacteria, is similar over the whole pile length, which might result from flowing water from the adjacent lake Mälaren which is relatively rich in oxygen. Many previous studies have shown that rot attacks decrease with the depth of the pile due to the reduction in oxygen. Other observations also show that the core of pine piles can be regarded as relatively resistant to attacks by bacteria.

See also **Berglund C, Holm G, Nilsson T & Bjordal C.** Durability of old timber piles: a case study. In: Butcher et al (eds) *Reuse of Foundations for Urban Sites: Proceedings of International Conference*, 2006. pp 79–86

12 Project Odin, Göteborg, Sweden

After the demolition of a 2-storey cast-in-situ concrete building with cellar that was built around 1940, a new 6–8-storey building was constructed. The old foundations consist of wooden friction piles in clay of a length of 20 m.

Two piles were dug out and a chemical analysis was conducted. In one of the piles, a 35 mm deep fungal attack ('pile fungus') was found just below the point of connection to the building. Another attack, by anaerobic bacteria, was found 750 mm below the point of connection to building. The other pile was intact. The effect of further attack on some of the piles was assessed with respect to their structural capacity. The geotechnical bearing capacity was assessed with respect to circumference of the pile group. The calculated pile load after the redevelopment was 125 kN. The total settlements calculated using pile loads, measured shear strength and in-situ vertical stress of between 50 mm and 80 mm.

13 New Acropolis Museum, Athens, Greece

The new Acropolis Museum, presently under construction, is being built on top of the early Christian part of the city going back to the 7th Century AD. The museum involves a typical case of foundation over existing ancient ruins and combines techniques of non- (or insignificantly) destructive foundation with reuse of existing footings in certain parts of the occupied area.

The entire installation will consist of nine structurally independent buildings founded on variable depths and having up to 4 underground and aboveground floors. The minimum practicable number of columns will be used, ie 43 for the main structure and 12 for the glass-covered access.

The damage to the ruins was estimated to be less than 0.2%, and was at locations selected to have antiquities of relatively minor importance. In the northwestern part of the area where the density of antiquities is high, foundations are on cast-in-situ concrete piles, either larger diameter single piles, or clusters of smaller diameter piles. In the central and southern part, the method of raft foundations will be employed.

One of the buildings intended for shops and offices, will be founded on the footings of an old residential building demolished to make room for the museum. Here, strip footings between the old existing footings will be applied. The restaurant will be founded on existing stone footings.

See also **Stamatopoulos AC, Stamatopoulos CA & Photiadis MG.** The new Acropolis Museum of Athens. In: Butcher et al (eds) *Reuse of Foundations for Urban Sites: Proceedings of International Conference*, 2006. pp 287–296

14 National Bank, Athens, Greece

The site of the new building is covered by the remains of the fortifications of the city of Athens constructed during the classical period and subsequently repaired a number of times. The old stone wall and gate are incorporated in the entrance of the building and exposed to open view. The ruins are reinforced with steel beams for protection against damage from earthquakes.

15 Reichstag, Berlin, Germany

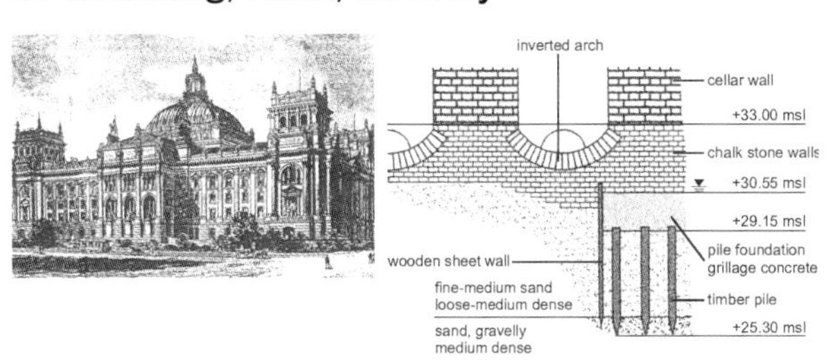

The construction of the 90 × 130 m 'Reichstag' building was carried out in 1884–1894. 3000 timber piles have been driven into the subsoil. Since the reunification of Germany, the 'Reichstag' building has been reconstructed.

The geometry of the foundations was examined by a number of trial pits and foundation core drillings. Core drillings, static penetration testing and driving tests were carried out in and next to the building. Due to the pre-loading and the driving in of the timber piles a medium dense packing was observed under the building, whereas the packing outside was rather loose. The timber piles underneath the building originally stood above the groundwater level but were exposed to a new environment when the groundwater was lowered for the construction of a metro line.

In the course of the reconstruction works, it was necessary to examine the timber piles in the area of the northern towers to estimate their future serviceability as they may have suffered from the change in environment. The bearing capacity of the timber piles and the surrounding soils were tested in 1996 in a test pit in the central area of the 'Reichstag'-building. 24 timber pile heads were uncovered from which 7 were used for test loading. One timber pile was used for a pulling test after the pressure test. The condition of the timber piles was tested in laboratory tests.

The splintwood was variously damaged due to bacterial attack, but the heartwood was mainly undamaged. As long as the piles remain constantly under water, their further use is possible. Due to the slow changes in the timber piles, a load transfer from the piles to the sand takes place. With the calculated creep, the long-time serviceability of the old foundation elements is guaranteed. The positive results from the various investigations carried out on the timber piles made it possible to formulate a plan for the foundations consisting of old and new foundation elements.

See also **Katzenbach R & Ramm H.** Reuse of historical foundations In: Butcher et al (eds) *Reuse of Foundations for Urban Sites: Proceedings of International Conference*, 2006. pp 395–404

16 Zürichhaus, Frankfurt am Main, Germany

One of the first office towers in Frankfurt am Main, the Zürich building was demolished in 2001. The basement of the building complex has been preserved for reuse in a future development.

17 Hessischer Landtag, Wiesbaden, Germany

The former plenary hall of the Hessian parliament was built in the 1960s. This building is situated in the city centre of Wiesbaden, an area which is well known for thermal springs. Due to the thermal groundwater conditions intrusions in the ground have to be reduced to a minimum. Therefore, the existing concrete-bored piles of the former building were to be reused as far as possible.

As the structure of the new building is quite different from the previous one, only a few piles were of interest regarding a potential reuse. After the demolition of the old building and the ground beams, the pile heads were exposed and inspected visually. During the integrity testing some pile heads were cut off due to damage during the demolition work, and were tested again. The load of the former building was carried by 81 piles of different diameters and lengths. 47 of these existing piles were located underneath the new building. 19 piles were intended for reuse and 17 could finally by incorporated into the new structure. Additionally, 73 new piles have been installed.

See also **Katzenbach R, Ramm H & Werner A.** Reuse of foundations in the reconstruction of the Hessian parliament complex: a case study. In: Butcher et al (eds) *Reuse of Foundations for Urban Sites: Proceedings of International Conference*, 2006. pp 385–394

18 Garden Towers, Frankfurt am Main, Germany

This 127 m high office building was finished in 1975. To modernise the building and to maximise the lettable space, the building was redeveloped in 2004. For this purpose, the entire cladding was removed, leaving only the structural work. To monitor the deformations of the subsoil and the adjacent buildings resulting from the load changes caused by the refurbishment works, geodetic measurements were carried out.

19 Das Silo, Hamburg, Germany

In the harbour of Hamburg, a former silo was converted into an office building. The entire foundation, the ground floor and four of the five silo tubes were preserved. Non-destructive testing was carried out. To distribute the changed building loads to the existing foundation which consists of concrete piles and a concrete ground beam grillage, a new slab was required in the centre of the new building.

20 Neuer Wall Arkaden, Hamburg, Germany

In the course of the refurbishment of the 6-storey retail and office building built in 1908, the existing timber piles were integrated in the new foundation layout. To assess the 8–9 m timber piles load test were carried out. In addition, the foundations were exposed and non-destructive testing, rotary core drilling and material tests were carried out. The load of the additional new storeys is transferred in new piles.

Appendix B: Financial risk case study

An idealised case study has been chosen to demonstrate the decision process for the selection of a suitable foundation solution. The case study consists of a redevelopment to provide an office development broadly based on the development outlined in Chapman & Marcetteau (2004). The case study is used for illustrative purposes and relies on a number of assumptions and simplifications.

B.1 Introduction

B.1.1 Description of redevelopment

The new development has the option to utilise existing foundations with a layout illustrated in Figure B.1. The existing foundation was used for a 5-floor office building with a concrete structure. The new planned superstructure is an additional storey taller but of a lighter steel construction and therefore has similar foundation loading. The new superstructure consists of:

- 5-storey high building plus single basement, ie 7 levels including the roof,
- Floor plan area: approx. 60 m × 60 m (3,333 m^2),

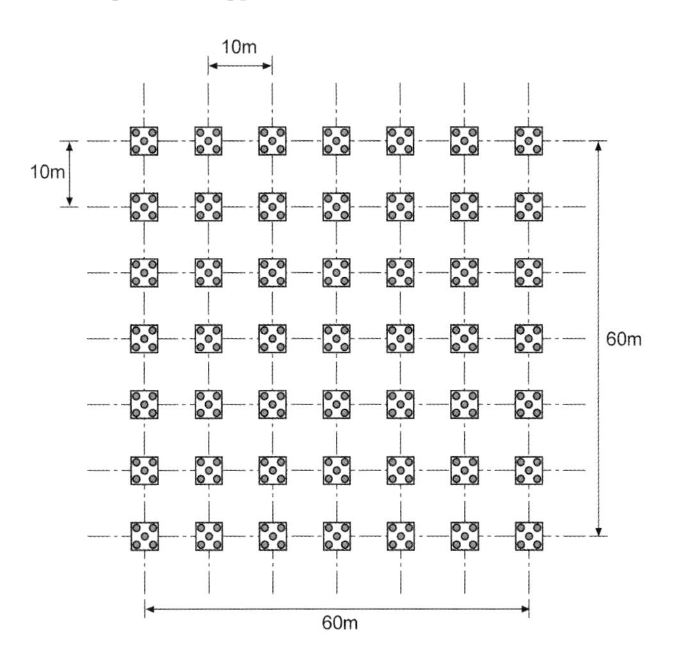

Figure B.1 Old pile layout of the idealised case study redevelopment (showing the 5-pile group design)

- Total floor space: 20,000 m^2,
- 10 m grid (7 columns × 7 columns = 49 columns); *note* that this grid has been regularised for the purposes of example and omits normal features of a building such as cores and structural stability systems,
- Dead load + live load = 12 kN/m^2, giving a typical central column load of 7,2000 kN.

The existing pile layout consists of 49 pile caps with 5 piles per cap, each supporting one column. The caps are centred in nodes in a regular grid at 10 m intervals (Figure B.1). The pile properties for each of the 245 piles in the 5-pile group configuration are:

- Diameter: 0.6 m,
- Pile length: 20 m,
- Working pile load: 7,200 kN/5 = 1,440 kN,
- Factor of Safety (FoS): 2, therefore the calculated ultimate load per pile is 2,880 kN.

A 5-pile group brings the benefit of redundancy: any individual pile can be omitted without bringing any other pile dangerously close to ultimate failure (provided it attains its calculated capacity) (see Cameron & Chapman 2004). This makes the building much more robust and less vulnerable to a defect in any single pile.

To consider the beneficial effect of the redundancy of a 5-pile group under each column, an alternative layout of a single pile per column location has also been considered in the analyses that follow. For this foundation solution the properties for each pile are:

- Diameter: 1.5 m,
- Pile length: 32 m,
- Working pile load: 7,200 kN,
- Factor of Safety (FoS): 2, therefore the calculated ultimate load per pile is 14,400 kN.

B.1.2 Foundation options

As presented in *Chapter 5*, several foundation options are possible when considering redeveloping a site. In the context of foundation reuse, the options can be extended to include:

- Complete replacement,
- Partial reuse,
  - full investigation of all reused piles,

- ❑ 'reasonable' investigation which will depend on the level of information available and the desired value of reuse load factor, R (as defined in *Chapter 3*),
- ❑ poor or limited investigation,

- Full reuse
  - ❑ full investigation of all reused piles
  - ❑ 'reasonable' investigation which will depend on the level of information available and desired value of reuse load factor, R,
  - ❑ poor or limited investigation.

B.1.3 Commentary on options

Each of the foundation options will be associated with different construction cost and programme, whole life cost, sustainability benefits and risk of foundation failure.

For the purpose of illustrating the financial implications of different foundation options, three principal foundation options have been considered and assessed using a risk-based decision analysis:

- Complete replacement,
- Complete reuse with 'reasonable' investigation,
- Complete reuse with 'poor or limited' investigation.

These three options have been considered to compare the costs and probability of failure of complete replacement with complete reuse and also to consider the implications of a less intense investigation that may fail to identify critical information on the existing foundations.

The extent of the investigation will depend on the level of caution that is appropriate for the proposed level of re-loading on the foundation as identified in *Chapters 3* and *5*. For the idealised case study it has been assumed that a 'reasonable' level of investigation relates to 10% of the pile population to be reused. In contrast, an investigation of only 3% of the pile population has been chosen to reflect a poor or limited investigation. For the 5-pile group foundations configuration this corresponds to investigating 25 piles and 7 piles, respectively. For the single pile per column configuration, it would correspond to checking 5 piles or potentially just one as a basis for decision making.

It has been assumed that the investigations comprised NDT testing only.

B.2 Risk-based decision model

The decision process for foundation reuse is presented in *Chapter 5* and the key stages to reduce the risk in foundation reuse are presented in *Chapter 3*. As part of the decision process, a comparative assessment of the various foundation options is recommended to identify the most appropriate or beneficial foundation options for the redevelopment based on the client's likely particular criteria:

- construction cost and programme,
- whole life cost,
- environmental impact (sustainability).

The potential financial risks and benefits associated with the quality of records for existing deep foundations and the expected status of the pile population for the three different foundation options for the idealised case study are evaluated and compared by means of risk-based decision analysis. This analysis is based on a Bayesian model (Jensen 2001) that has been developed for an idealised redevelopment scheme to compare the total expected cost of several foundation options, including complete reuse. The model also compares the likelihood of failure of the structure caused by the foundations with standard levels of acceptable risk. While this sort of modelling may not be suited to individual redevelopment projects, it provides a useful tool for undertaking this comparison study.

B.2.1 Risk-based decision analysis model

The risk-based decision analysis model used to compare the different foundation options as part of the idealised case study is based on Bayesian networks or influence diagrams. In this quantitative model, the risks are formulated as the product of the probability of failure and the corresponding financial consequences. The decision criterion is two-fold:

- minimise the total expected cost,
- satisfy minimal acceptable levels of probabilities of failure.

The influence diagram built for this case study is shown in Figure B.2.

The influence diagram includes:

- a single decision node (rectangle) representing the decision alternative for the foundation option,

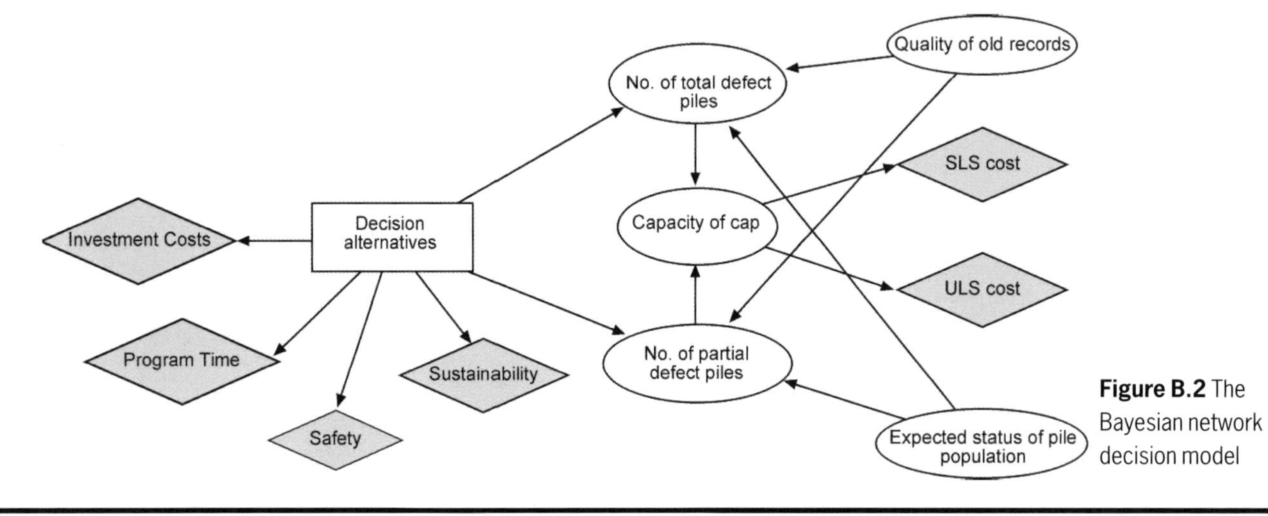

Figure B.2 The Bayesian network decision model

- five chance nodes (ovals) representing probabilistic variables, and
- six utility (or cost) nodes (rhomboids) representing different costs associated with the different foundation alternatives.

The total expected cost for each of the foundation alternatives is the sum of all of the cost nodes.

Two of the cost items, cost of serviceability limit state (SLS) failure and cost of ultimate limit state (ULS) failure depend on conditioning provided by the model for the probability of the number of defect piles. The probabilities of the number of totally defective piles and the number of partially defective piles are conditioned on three nodes representing the quality of old records, expected status of the pile population and the foundation alternative being considered. Further information on the conditional probabilities in the model is provided in *Section B.4*.

Table B.1 gives an explanation and a summary of the nodes, together with the different states of the nodes used in the decision model. Further information on each of these nodes is provided in the chapter references given in the Table.

The influence diagram is built and solved with the software Hugin Expert (Jensen et al 2002). Further information on the theory and further references on influence diagrams are given later.

Prior knowledge of the piles (eg expert judgements, knowledge based on experiences and site-specific documentation) was considered important input into the decision models.

Initially, the results presented are *forward predictions* (ie based on the stage before investigations are carried out. Observations from various non-destructive testing (NDT) methods can form an important extension of the decision model. Influence diagrams can be used to reason under uncertainty and to update existing information with observations from site investigations, and are described in *Section B.4*.

B.2.2 Assumptions relating to cost

For the purposes of this case study the total expected cost (TEC) for the different foundation solutions will consist of:

- Base costs (total investment costs),
- Financial benefits,
- Cost of failure.

B.2.2.1 Substructure base costs

The investment costs given in Table 4.2 have been estimated based on the article by Chapman & Marcetteau (2004). However, they have been modified to account for the differences in the foundation options and the relatively congested nature of the site.

B.2.2.2 Financial benefits of foundation reuse

Financial benefits associated with foundation reuse have been considered including:

- *Construction programme*, ie how early the building is completed. As identified in *Chapters 2, 4* and *5*, foundation reuse has the potential to reduce the construction programme as the scope of the foundation works are reduced or, in the case of complete reuse, are removed from the programme.
- *Safety during construction.* Although not considered 'unsafe', piling activities do have certain safety risks

Table B.1 Summary of nodes in the decision model

Type of node	Name of node	Number of states the node
Decision node (Rectangle)	Reuse alternatives	3 Complete replacement Full reuse with reasonable investigation Full reuse with limited investigation
Chance node (Oval)	Number of totally defective piles in the pile population	8 (0 to 7 totally defective piles)
	Number of partially defective piles in the pile population	11 (0 to 10 partially defective piles)
	Capacity of substructure	3 (Sufficient, SLS, ULS)
	Quality of the old records	2 (Good, Bad)
	Expected status of the pile population	3 (Good, Average, Bad)
Utility node (Rhomboid)	Investment costs	—
	Program time	—
	Construction safety	—
	Sustainability	—
	Costs for SLS failure	—
	Costs for ULS failure	—

Table B.2 Investment costs for the substructure, based on Chapman & Marcetteau (2004)

	Complete replacement	Full reuse with reasonable investigation	Full reuse with limited investigation
Desk study	£5,000	£5,000	£5,000
Ground investigation, including interpretation	£35,000	£35,000	£35,000
Sample testing of the pile population, including assessment report	£0	£75,000	£15,000
Piles and caps	£1,320,000	£120,000	£120,000
Basement structure	£2,250,000	£2,200,000	£2,200,000
Total substructure investment cost	**£3,610,000**	**£2,435,000**	**£2,375,000**

associated with them and therefore the reduction or elimination of piling activities could have a beneficial effect on the construction safety of the works.

- *Sustainability issues,* ie reduction in embodied energy and not adding to congestion in the ground. A measurable indicator of sustainability is the overall energy inputs required to construct new foundations and where necessary to remove obstructions in the ground at new foundation locations. The reuse of existing foundations will provide sustainability in terms of future development of the site by not adding to the congestion in the ground. See *Chapters 2, 4* and *5*.

Realistic and extreme benefit costs as given in Table B.3 have been chosen to demonstrate their influence on the model.

B.2.2.3 Price estimation of the consequences of foundation failure

If an option to reuse old piles is considered, the quality of the old piles may strongly affect the risk that the new structure will not fulfil the specified requirements defined by ultimate limit state (ULS) and serviceability limit state (SLS) failure conditions as outlined in *Chapter 3*. In this case study, it was assumed that the consequences of failure conditions apply exclusively to structural damage and delays due to interruption in occupation of the building during the repair phase. The assumptions used for estimating the costs for SLS and ULS failure for the idealised case are given in Table B.4.

The estimates of the ULS and SLS failure costs do not include the following:

- human consequences of the failure, apart from the direct compensation payment,
- legal costs,
- cost of negative publicity and possible loss of future potential contracts,
- fines or penalties by safety authorities etc.,
- loss of 'no claims' on future company insurance policies,
- management costs in resolving issues, or
- overall disruption to company operations from loss of a critical facility.

These items have not been included in the idealised case study due to lack of available data for assessing those costs and to create a simplified comparison between the different options. However, for a real case, *these items can be*

Table B.4 Assumptions for estimating the costs for SLS and ULS failure

Item	ULS failure costs	SLS failure costs
Company not working	£2.6 million (Based on 1 month of non-productive staff with an annual wage bill of £31,250,000)	£0.58 million (Based on 1 week of non-productive staff with an annual wage bill of £31,250,000)
Disruption and relocation costs	£8.0 million (Based on 6 months × £1.3 million/month)	£0.27 million (Loss of amenity)
Repair cost	£3.0 million	£0.5 million
Compensation to individuals	£0.5 million for personal injury	No injury
Total	**£14.1 million**	**£1.35 million**

Table B.3 Assumptions for benefits to include in the decision model

Scenario	Benefits	Complete replacement	Full reuse with reasonable investigation	Full reuse with limited investigation
Benefits assumption 1: realistic	Construction programme	£0	£200,000	£200,000
	Safety	£0	£100,000	£100,000
	Sustainability	£0	£300,000	£300,000
	Total	**£0**	**£600,000**	**£600,000**
Benefits assumption 2: extreme	Construction programme	£0	£800,000	£800,000
	Safety	£0	£400,000	£400,000
	Sustainability	£0	£1,000,000	£1,000,000
	Total	**£0**	**£2,200,000**	**£2,200,000**

significant and represent real costs which should be considered.

B.2.3 Estimation of number of defective piles in pile population

The presence of defects in the piles will affect the performance of the new structure and depending on the number, size and location of the defective piles around the structure and individual pile caps may result in SLS or ULS failure of the structure.

In this case study, defective piles have been classed as:

- *Non-defective pile:* < 25% loss of load-carrying capacity assumed not to affect the performance of the structure.
- *Partially defective pile:* 25–75% loss of total load-carrying capacity.
- *Totally defective pile:* > 75% loss of total load-carrying capacity.

Estimates of the *absolute maximum* number of totally and partially defective piles that can be present in the total pile population of 245 at the site in this particular case study have been made based on the definitions of the defects and reference to previous studies, see eg Cameron & Chapman (2004). Information on the frequency of defective-bored piles, identified through NDT can be found in:

- Davis & Dunn (1974) who report 9.7% out of a total of 717 piles tested on 5 projects;
- Fleming et al (1985) found 1.5% defective out of a total of 5000 piles tested and 1.9% defective out of a further 4450 piles tested;
- Ellway (1987) reports 4.2% defective piles out of a total of 4400 piles tested;
- Thasnanipan et al (1988) state 3.3% defective of a total 8689 piles tested;
- Lew et al (2002) report 7% defective within a population of 380 piles and 1.5% defective of a total piles tested;
- Preiss & Shapiro (1981) suggest that approximately 5–10% of the piles on a project could be defective.

The estimated absolute maximum numbers of defective piles for the idealised case study of 245 piles are presented in Table B.5.

B.2.4 Estimations of probabilities of defective piles

The number of defective pile presented in Table B.5 represents the maximum number of defective piles. However, these can be treated as upper-bound values and the number of defective piles present in the pile population can vary between zero and these upper-bound values.

The probabilities for the number of totally and partially defective piles must therefore be estimated. The estimate is site-specific and primarily based on:

- the quality of old records,
- the expected status of the pile population, and
- any investigations made at the site.

In the case study, it is assumed that any piles that are *known* to be either partially or totally defective do not constitute any problem: these piles are not incorporated into the new structure and are replaced by new piles. Thus, only the *unknown* defective piles are of concern.

For the case study, the estimation of the probability for the number of unknown defective piles has been conditioned upon the quality of the old records and the expected status of the pile population alone. *The influence of the results of any investigations on the site is not included for simplicity* and also to provide a clear indication of the effects of these two different preconditions to be illustrated. However, the risk-based decision model can be extended to include the results of pile testing by means of an additional influence diagram as presented in *Section B.4*.

Quality of old records

Two states have been chosen for the *quality of old records* in the case study: 'good' or 'bad'. From good quality old records it should be possible to identify piles that are likely to include a defect and therefore reduce the number of unknown defective piles. Bad records are unlikely to provide indications of defective piles and therefore it will be less easy to identify all of the defective piles in the population from these records. Thus, the quality of old records will influence the expected number of defective piles in the population. This illustrates the benefits of keeping good records to facilitate future reuse of piles as advocated in *Chapter 7*.

A state of 'no knowledge' for the quality of old records has been included in the model as 50% 'good' and 50% 'bad'.

Expected status of the pile population

The expected status of pile population is a function of:

- the care taken when forming the initial piles (based on the perceived quality of the piling contractor);
- any corrosion or loss in durability that has taken place over the life of the piles, and
- the care taken when the first superstructure was demolished.

In the idealised case study, the expected status of pile population is can be in either one of three states: 'good', 'average' or 'bad'. The number of (unknown) defective piles depends on the expected status of pile population, and there are likely to be fewer defective piles if the expected status of pile population is 'good' than if it is 'bad'. The expected status will change as investigations of the foundations are undertaken on the site or demolition is completed.

Table B.5 Assumed maximum number of defective piles

Type of defect	Assumed maximum number of defective piles (245 piles in 5-pile group configuration)	
	Number	%
Partially defective pile	10	4.1
Totally defective pile	7	2.9
Partial and total defect pile	17	6.9

The results also consider a state of 'no knowledge' for the quality of old records. This has been included in the model as 33% 'good', 33% 'average' and 33% 'bad'.

Probabilities of defective piles

In the idealised case study, the maximum number of defective piles and the probability of finding defective piles have been estimated based on literature and reasoning. For the case study, the probability of finding zero defects is high, approximately 0.56 to 0.998 depending on the input conditions as shown in Figure B.3. It was also assumed that this probability would decrease exponentially for each consecutive increment in the number of defective piles to be found.

To estimate the probability of finding no defective piles, one could ask, eg: 'What is the probability of finding at least one pile that has a loss of total capacity greater than 75% at this particular site with no investigations to support the estimate?'. Figure B.3 shows the corresponding probabilities for each number of defective piles given the different assumed states.

The estimate of the number of defective piles in the population will also depend on the percentage of the piles in the population that are tested. The estimates given in Figure B.3 represent the pre-investigation stage. The results of different testing regimes can be incorporated into the risk-based decision analysis demonstrated in *Section B.4*.

B.2.5 Capacity of substructure and effects of increased redundancy offered by multi-pile groups

The reliability of a piled foundation (eg its ability to perform as intended under operating conditions) is not only affected by the construction quality of each pile in the population, but also by the layout of the piling system (Cameron & Chapman 2004). The consequence of a defect in any individual pile depends on the redundancy offered by the piling layout. Larger pile groups are inherently more redundant than smaller groups or single pile groups due to the reserve capacity offered by neighbouring piles and the ability to redistribute load through the soil and pile cap.

The assumed effect of redundancy, given the definitions of partially and totally defective piles, for a five-pile cap design is given in Table B.6. It should be noted that no consideration was made as to which position the defective pile occupies within the pile cap. For comparison, the assumed effect of redundancy for a single pile cap design is also given in Table B.6. These assumptions were used to calculate the probability of SLS and ULS failure for the superstructure given the uncertainty of the number of partially and totally defective piles. Further details of the mathematics for calculating the probability for each of these combinations to be present, given different numbers of defective piles in the whole pile population are given in *Section B.4*.

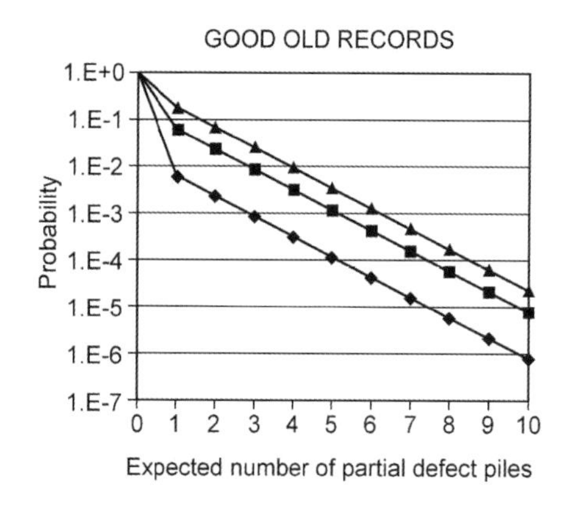

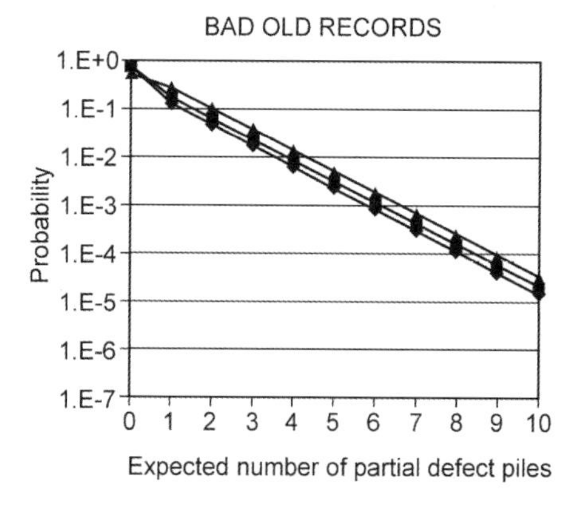

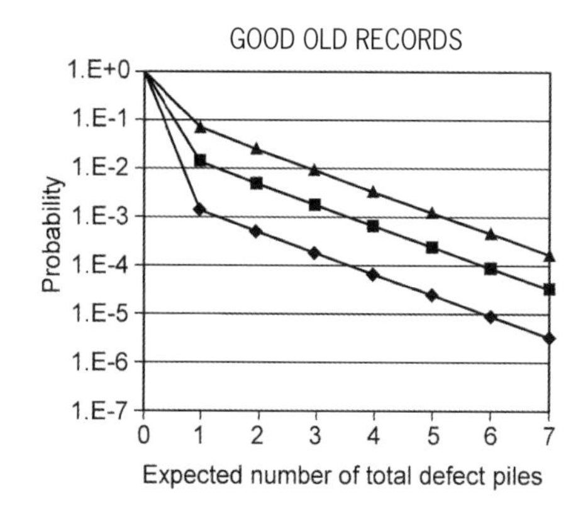

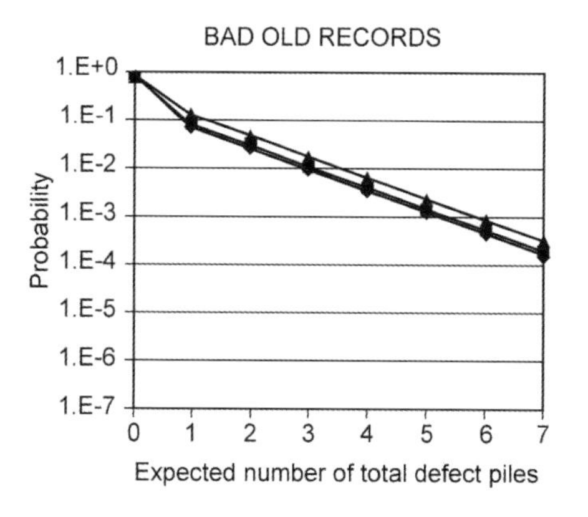

Expected status of pile population is GOOD
Expected status of pile population is AVERAGE
Expected status of pile population is BAD

Figure B.3 Estimated probabilities of the number of partially and totally defective piles for the idealised case study

Table B.6 Effect of foundation redundancy on superstructure failure

	No superstructure failure	SLS failure	ULS failure
5-pile group	≤ 1 partially defective pile	2 partially defective piles **or** 1 totally defective pile	≥ 3 partially defective piles **or** ≥ 2 totally defective piles **or** 1 partially defective pile and 1 totally defective pile
Single pile group	0 defective piles	1 partially defective pile	1 totally defective pile

Other aspects that will improve the ability of the superstructure to accommodate defective piles without reaching SLS failure and ULS failure include:

- redundancy of the substructure and superstructure to spread any load-carrying capacity deficit to non-defective foundations,
- increased caution in re-loading through a reduction in the reuse load factor, R.

For the purposes of this idealised case study, any beneficial effects of the remainder of the building structure have been ignored and a reuse load factor, R, of unity has been assumed based on the similarity between the new and old developments.

B.2.6 Optimum reuse solution for different cases

Three foundation options have been included in the decision analysis:

- complete replacement,
- complete reuse with reasonable investigation,
- complete reuse with poor or limited investigation.

These have been considered with three sets of prior knowledge of the quality of the old records (eg foundation design basis and as-built drawings) and the expected status of the pile population (eg anticipated material deterioration and care taken during demolition of the superstructure):

- Good,
- Bad,
- No knowledge.

Three separate analyses have been carried out.

- 5-pile group: a base case to compare the alternatives generally
- 5-pile group including financial benefits: a study to demonstrate the effects of differing financial benefits
- Single pile group: a study to demonstrate the effects of redundancy in the foundations

The results of the decision analysis are given in Figures B.4–B.6 and include:

- total investment costs or base costs,
- total expected cost for SLS and ULS failure,
- total expected cost, including the costs of SLS or ULS failure and any financial benefits,
- probability of SLS and ULS failure.

The results represent the stage prior to investigations being carried out (ie they are forward predictions). The results of an investigation can be incorporated into later stages of modelling (see *Section B.4*).

The result of the decision analysis, ie the total expected cost (TEC) of each decision alternative is presented as a sensitivity analysis in Figures B.4–B.6. The sensitivity analysis is an important part of any decision analysis for investigating which variables will influence the result most. Identifying the most sensitive variable, forces us to investigate how certain we are of the estimated value of these variables.

The acceptable probabilities of SLS and ULS failure (in accordance with BS EN 1990 for a 50-year design life as identified in *Chapter 3*) are included in the figures to demonstrate which of the options provide a risk of failure that satisfies the requirements of BS EN 1990.

Five-pile group excluding benefits

Figure B.4 presents the result of the analysis of the 5-pile group design excluding any financial benefits.

As a general trend, the results indicate that the total expected cost of SLS or ULS failure is low and therefore the TEC is equivalent to the total investment cost for all decision alternatives. This is in part due to the fact that the costs for ULS and SLS failure do not include some costs (see *Section B.2.2*) and are multiplied with the probabilities of ULS and SLS failure, respectively, which are exceedingly small for this scenario.

The effects of the quality of the old record and the expected status of the pile population can be seen in the calculated probabilities of SLS and ULS failure. The probability of failure increases as the records and expected status decrease.

As a result of the small value of the TEC for SLS or ULS failure, the TEC for the foundation options are virtually independent of the quality of the old records or the expected status of the pile population.

The results show that the decision alternative with the lowest TEC is 'complete reuse with limited investigations'. However, the probability of failure for this foundation option does not satisfy the requirement of BS EN 1990. In fact, for the case-specific assumptions in this idealised case study the only foundation option that fulfils the probability requirements for failure is 'complete replacement'. This result is dependent on the assumptions made, in particular the maximum number of defective piles, the probability of

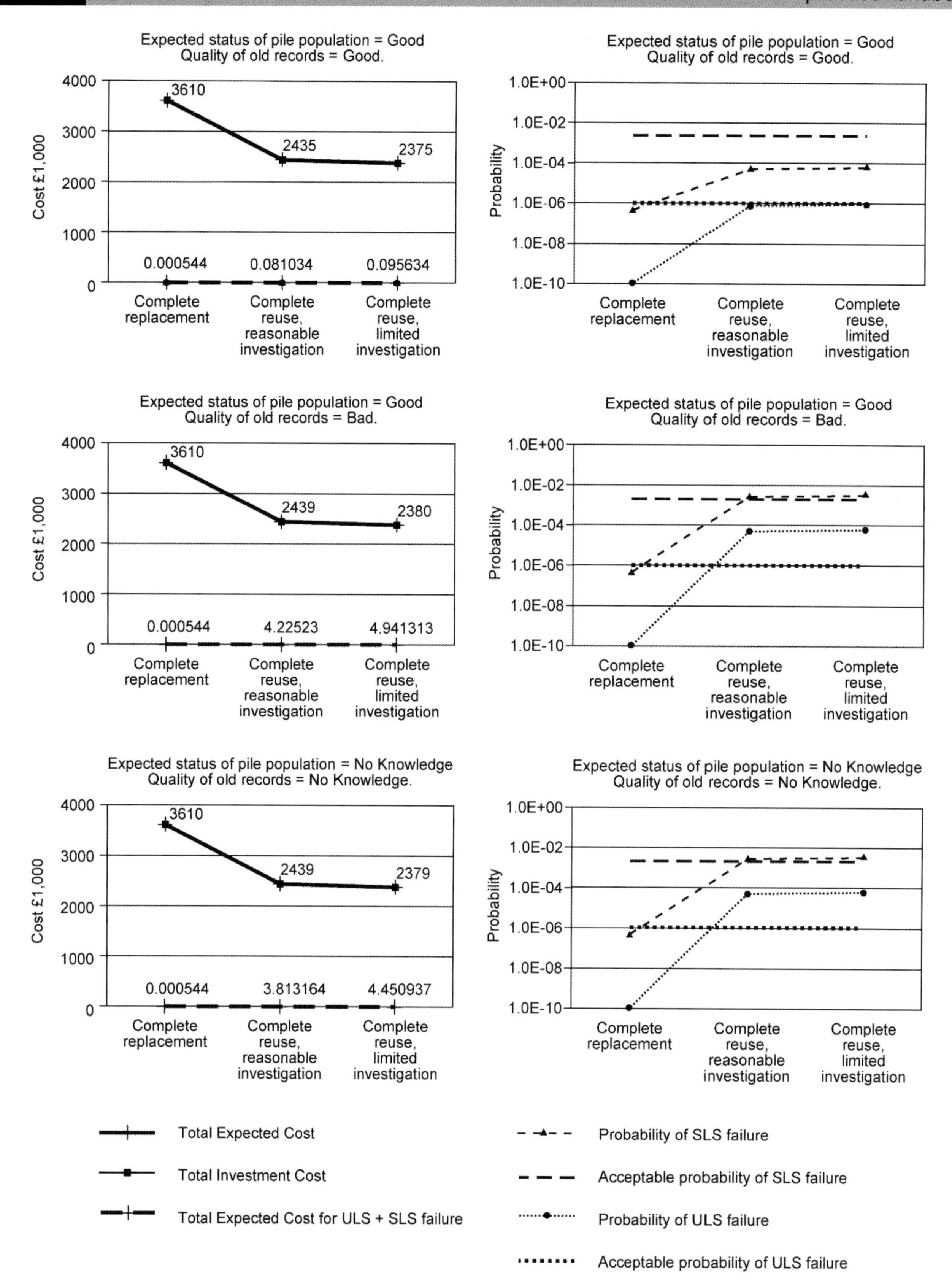

Figure B.4 Results from the Bayesian model for a 5-pile group

defective piles and the effect of foundation redundancy on superstructure failure as defined earlier. It does not conclude that foundation reuse will be incompatible with the probability of failure requirements of BS EN 1990 for all cases.

5-pile group including benefits

Figure B.5 presents the result of the same 5-pile group considered but includes the effects of different financial benefits. As the same prior information is used in the analysis, therefore the probabilities of failure are unchanged and the comments made previously on the probabilities of failure are relevant.

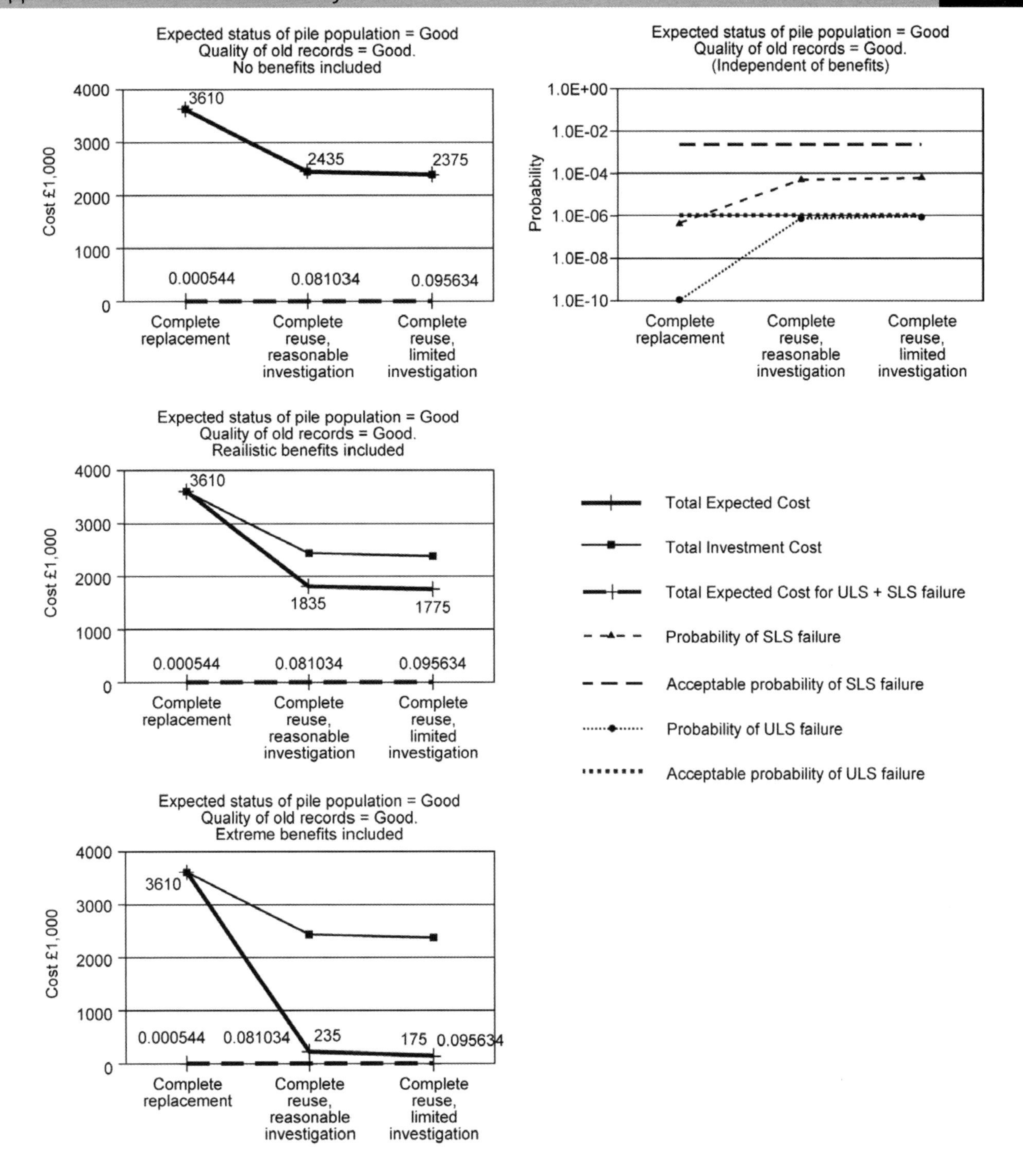

Figure B.5 Results from the Bayesian model for a 5-pile group including financial benefits

The magnitude of the benefits has a significant effect on the TEC of the building substructure. The extreme level of benefits considered in the idealised caste study virtually balances the investment costs. In practice, costs associated with benefits will be difficult to quantify, however, this comparison shows the significant impact they can have on the feasibility of the different options and therefore the importance of their consideration in the decision process.

Single pile group

The results from the Bayesian model for a 1-pile group design are shown in Figure B.6.

The lack of redundancy in a single-pile group is clearly indicated by the relatively high probabilities of failure for foundation reuse predicted by the model (> 10^{-2} in comparison with < 10^{-4} for the 5-pile group). These high probabilities are reflected in the TEC of ULS and SLS failure which is now a significant contributor to the TEC, in particular for the no knowledge prior condition. The probabilities of failure for foundation reuse do not satisfy the requirements of BS EN 1990 and again, complete replacement is the only compliant option.

The influence on the quality of prior information on the probability of ULS failure can be clearly seen (> 10^{-2} for good prior information and > 10^{-1} for no knowledge of prior information). For good prior information the probabilities of failure are sufficiently small for foundation reuse to look economically favourably. However, for no knowledge of the

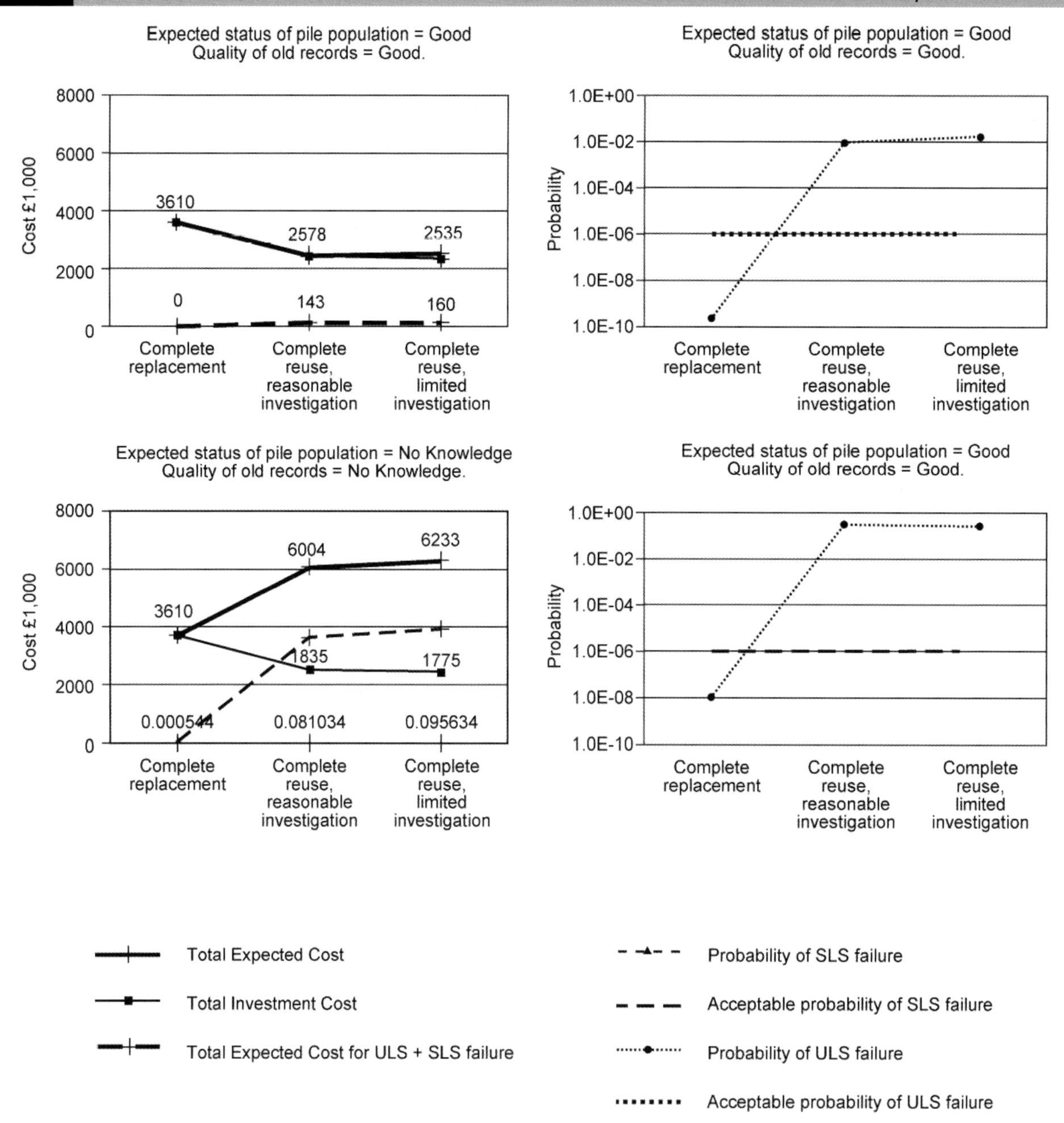

Figure B.6 Results from the Bayesian model for a single pile group

prior condition, the TEC for foundation reuse is significantly higher than complete replacement.

B.2.7 Discussion and conclusions

As can be seen from Figures B.4–B.6, the results of the decision analysis depend on the prior information, the benefits and the piling layout.

Status of prior information

The benefits of better prior knowledge on the expected status of the pile population (from quality of construction, care in its demolition and defects introduced over their lifetime) and the quality of the old records (giving certainty of what was actually installed) have been demonstrated through reduced TEC and probabilities of failure predicted by the case study model.

Structural redundancy

The effects of improved structural redundancy of multiple pile groups has been demonstrated by significantly reducing the probabilities of failure for the more robust 5-pile groupings. For a 5-pile cap design, the TEC of structural failure was shown to be generally low compared with the investment costs due to the small probabilities of failure (10^{-4} –10^{-6}). In contrast, the less robust or less redundant single pile group was shown to have more significant probabilities of failure (10^{-2}–10^{-1}) and hence larger total expected costs of structural failure for foundation reuse. It should be noted that there are several risk costs associated with SLS and ULS failure that were not included in the decision model such as legal costs, human injuries or costs for negative publicity. If those costs were included as well, the TEC for ULS and SLS failure could rise significantly. However, in order for the failure cost to have an effect on the TEC, the size of the failure cost must be of the order of 10^4–10^6 times larger than the investment costs.

Benefits of investigation
The results from this stage (before investigation is actually carried out) indicate that the alternative 'complete reuse with limited investigations' generally produces a lower TEC than the alternative 'complete reuse with reasonable investigations'. The input data used in the case study implies that the expected additional data from the additional level of investigation does not lower the expected total cost of failure more than the cost of obtaining that data. However, the level of investigations will influence the probability of failure and therefore higher levels of investigation are more likely to reach the acceptable probability of failure.

In practice, the choice of investigation will not be dependent on an analysis such as this risk-based decision model but instead will be a result of engineers exercising professional skill and care through engineering judgment for the degree of re-loading and an appropriate level of caution as identified in *Chapters 3* and *5*.

Variability of investment costs
In the idealised case, the investment costs were not varied. In reality, the investment costs will not be known accurately until construction has been completed and therefore investment costs would be more appropriately represented by a range or distribution of costs. The size of the investment cost is of great importance to the result of the decision analysis and should be investigated further.

Dominance of probability of failure over TEC
For the case study considered, it is the magnitude of the probability of a structural failure that will dominate the decision model (ie whether it will exceed or be lower than the acceptable probability of failure). The acceptable probability of failure is a regulated probability, and not a probability that the site owner, or the decision-maker, can choose randomly.

Suitability of foundation reuse
The idealised case study presented demonstrates the financial benefits that foundation reuse can present to a project in particular where there is redundancy in the foundation system to reduce the reliance on each and every foundation.

The foundation reuse options presented in the case study do not meet the requirements of BS EN 1990 for probability of failure. This is based on the site-specific conditions such as the assessment of the maximum number of defect piles and the probability of defective piles. It does not conclude that foundation reuse will be incompatible with the probability of failure requirements of BS EN 1990 for all cases.

In practice, complete reuse with a reuse load factor of unity will seldom be adopted as the requirements of the new and old developments are unlikely to be so similar. It is likely that some level of caution will be adopted through a reuse load factor below unity and supplemental foundations will be required to satisfy the specific loading distribution of the new structure. These features in conjunction with consideration of any redundancy offered by the building structure may reduce the assessed probabilities of SLS and ULS failure.

TEC *vs* potential financial exposure
It should be noted that the TEC does not represent a client's potential financial exposure on a project. The TEC factors the failure costs by their probability of occurrence. However, in reality, the costs will either occur in their entirety or not at all. Despite this, consideration of TEC is a valuable tool to allow a balanced view of the potential risk of failure to be taken when considering the different options.

B.3 The capacity of the substructure (5 piles/cap)

The capacity of the substructure is a vital component of the decision model for the idealised case study since it directly reflects the potential for SLS and ULS failure. In the case study, the definition of SLS and ULS was devised according to the rationale outlined in *Chapter 3* and *Table B.6*.

Now, assume that we either know or can estimate probabilities of the existence of a certain number total and partially defective piles in the population. What is the lowest probability that defect piles will form a configuration within a single cap that will result in classifying it as SLS or ULS failure, or a sufficient capacity state? These three states are mutually exclusive. Further, we assume that the defect piles are spatially independent, ie randomly positioned. According to the assumptions made in this case study, the expected maximum number of defect piles in the populations was considered to be 7 and 10 for total and partially defective piles, respectively.

The problem to be solved can be exemplified by the following. Suppose that there are zero totally defective and three partially defective piles in the whole pile population. What is the probability that those existing defect piles form any of configuration(s) that will lead to either of three states: sufficient capacity state, SLS or ULS failure? That is, how probable is it that a random process will produce, eg:

(1) two partially defective piles within the same cap, thus produce an SLS-failure,
(2) three partially defective piles within the same cap, thus produce an ULS-failure, or
(3) one partially defective pile, thus produce a sufficient capacity of the substructure?

Table B.7 Effect of redundancy for a 5-pile cap

No. of partially defective piles in cap	No. of totally defective piles in cap	Type of superstructure failure
0	0	None
1	0	None
2	0	SLS-failure
≥3	0	ULS-failure
0	1	SLS-failure
0	≥2	ULS-failure
≥1	≥1	ULS-failure

These probabilities must be calculated for any given combination of the number of total and partially defective piles, from zero to 7 and 10, respectively.

To calculate the probability of each SLS and ULS failure, a multinominal probability density function was invoked as it was considered best suited to describe the problem. Below follows a brief description of the main steps included in the calculation procedure.

Definitions

Each pile can belong to one of three stages:

- $k = 1$, no defect,
- $k = 2$, partially defective, or
- $k = 3$, totally defective.

There are $N = 245$ piles in total, 49 caps, and 5 piles per cap.

Let X_k be a number of piles in a cap that belongs to state k. For example, $X_1 = 5$ means that in a cap there are 5 piles with no defect.

Let Y_k be a number of piles among 245 that belong to state k. For example, Y3 = 7 means that among 245 there are 7 piles that belong to state $k = 3$, ie totally defective.

Further, let p_k be probability that a pile belongs to state k. All piles are independent of each other.

Three events are defined:

- A (sufficient capacity) $= \{Y_2 \leq 1\} \cap \{Y_3 = 0\}$
- B (SLS failure) $= \{X_2 = 2\} \cup \{X_3 = 1\}$
- C (ULS failure) $= \{X_3 \geq 2\} \cup \{X_2 \geq 1, X_3 \geq 1\} \cup \{X_2 \geq 3\}$

Problem formulation

To find the probabilities P(A), P(B) and P(C), the probabilities p_k, where $k = 1, 2,$ and 3 (ie p_1, p_2 and p_3), must be calculated.

Solution

$P(A) = P(Y_2 \leq 1, Y_3 = 0) = P(Y_2 = 0, Y_3 = 0) + P(Y_2 = 1, Y_3 = 0)$

$P(B) = P(\{X_2 = 2\} \cup \{X_3 = 1\}) = P(X_2 = 2) + P(X_3 = 1) - P(X_2 = 2, X_3 = 1)$

$P(C) = P(\{X_3 \geq 2\} » \{X_2 \geq 1, X_3 \geq 1\} \cup \{X_2 \geq 3\}) = 1 - P(\{X_3 \leq 1\} \cap \{\{X_2 = 0\} \cup \{X_3 = 0\}\} \cap \{X_2 \leq 2\} = 1 - P(X_2 = 0, X_3 = 1) - P(X_2 = 0, X_3 = 0) - P(X_2 = 1, X_3 = 0) - P(X_2 = 2, X_3 = 0)$

The vector (Y_1, Y_2, Y_3) is distributed according to a multinominal probability density function with parameters: N, p_1, p_2, and p_3. The complete depiction of the computation procedure is beyond the scope of this document so interested readers are referred to Rice (1995). The result, ie the final formulas for calculating probabilities of cap capacity are presented below.

$P(A = \text{sufficient capacity}) = p_1^{245} + 245p_1^{244}p_2$

$P(B = SLS) = 10p_2^2(1-p_2)3 + 5p_3(1-p_3)^4 - 30p_1^2p_2^2p_3$

$P(C = ULS) = 1 - 5p_1^4p_3 - p_1^5 - 5p_1^4p_2 - 10p_1^3p_2^2$

$p_1 + p_2 + p_3 = 1$

Table B.8 presents the calculated probabilities for:

(1) sufficient capacity state,
(2) SLS failures, and
(3) ULS failures.

P(Sufficient) is equal to P(A), P(SLS failure) is equal to P(B), and P(ULS failure) is equal to P(C). The probabilities are calculated for all possible combinations of total and partially defective piles in the pile population.

B.4 The number of unknown defect piles for each decision alternative

To calculate the probabilities for each given number of unknown total and partially defective piles to be present in the pile population, a testing model (TM) was built parallel to the main decision model. The TM was built to generate input data to the decision model used in the idealised case study. Similar to the main decision model built for the idealised case, TM is rather simple in its structure and constrained by several assumptions.

One of the features of the TM is to illustrate how information or data from NDT-tests on the piles and field investigations can be incorporated into a decision-making process. It is incorporated by means of Bayesian updating. Bayesian statistics differs from classical statistics in that it includes all kinds of data, ie both objective (hard data) and subjective (soft data) information for making a prior estimate of the probability of a certain event. In fact, it requires a prior belief. By using Bayes' theorem, the prior estimate is updated to posterior probabilities (Vose 1996):

$$P(A_i|B) = \frac{P(B|A_i) \cdot P(A_i)}{\sum_{i=1}^{n} P(B|A_i)P(A_i)}$$

where A_i and B are events, $P(A_i)$ and $P(B)$ are the probabilities of A_i and B, respectively, $P(B|A_i)$ is the probability of B given A_i, and $P(A_i|B)$ is the probability of A_i given B. A_i represents the prior information and B is the new information used for updating the prior information. The more hard data that are used to update the prior estimate, the more the updated information will reflect the collected hard data. The prior estimate may be solely based on subjective information (ie expert judgement).

The testing model is built by means of an influence diagram. Influence diagrams consist of a directed acyclic graph over chance nodes (ovals), decision nodes (rectangles) and utility nodes (rhomboids), with the following structural properties: a directed path comprising all decision nodes (ie there is a temporal sequence of decisions), and the utility nodes have no children. Quantitative specifications require that:

(1) decision and chance nodes have a finite set of mutually exclusive states,
(2) the utility nodes have no states,
(3) a conditional probability table $P[A|pa(A)]$ is attached to each chance node A, and finally,
(4) a real-valued function over $pa(U)$ is attached to each utility node U.

By introducing evidence in an influence diagram, prior probabilities can be updated to posterior probabilities, as in

Table B.8 Calculated probabilities for each state: sufficient, SLS failure or ULS failure for all possible combinations of partially and totally defective piles. No_TD means the number of totally defective piles and No_PD means the number of partially defective piles

No_TD						**0**					
No_PD	**0**	**1**	**2**	**3**	**4**	**5**	**6**	**7**	**8**	**9**	**10**
Sufficient	1	1	0.99935	0.99854	0.99742	0.99600	0.99429	0.99229	0.99002	0.98747	0.98466
SLS-failure	0	0	0.00065	0.00145	0.00254	0.00392	0.00557	0.00748	0.00965	0.01206	0.01470
ULS-failure	0	0	0	0.00002	0.00004	0.00008	0.00014	0.00022	0.00033	0.00047	0.00064

Cont...

No_TD						**1**					
No_PD	**0**	**1**	**2**	**3**	**4**	**5**	**6**	**7**	**8**	**9**	**10**
Sufficient	0.97992	0.97927	0.97846	0.97734	0.97593	0.97421	0.97222	0.96994	0.96739	0.96458	0.96151
SLS-failure	0.02008	0.02024	0.02072	0.02150	0.02258	0.02394	0.02558	0.02747	0.02961	0.03199	0.03459
ULS-failure	0	0.00049	0.00082	0.00115	0.00149	0.00184	0.00221	0.00259	0.00300	0.00343	0.00389

Cont...

No_TD						**2**					
No_PD	**0**	**1**	**2**	**3**	**4**	**5**	**6**	**7**	**8**	**9**	**10**
Sufficient	0.95985	0.95904	0.95792	0.95650	0.95479	0.95279	0.95052	0.94797	0.94516	0.94209	0.93877
SLS-failure	0.03950	0.03966	0.04013	0.04091	0.04197	0.04332	0.04493	0.04680	0.04891	0.05126	0.05383
ULS-failure	0.00066	0.00130	0.00195	0.00259	0.00323	0.00389	0.00455	0.00523	0.00593	0.00665	0.00740

Cont...

No_TD						**3**					
No_PD	**0**	**1**	**2**	**3**	**4**	**5**	**6**	**7**	**8**	**9**	**10**
Sufficient	0.94026	0.93914	0.93772	0.93601	0.93401	0.93174	0.92919	0.92638	0.92331	0.91999	0.91643
SLS-failure	0.05828	0.05844	0.05891	0.05967	0.06073	0.06205	0.06364	0.06549	0.06757	0.06989	0.07243
ULS-failure	0.00146	0.00242	0.00337	0.00432	0.00526	0.00621	0.00717	0.00814	0.00912	0.01012	0.01114

Cont...

No_TD						**4**					
No_PD	**0**	**1**	**2**	**3**	**4**	**5**	**6**	**7**	**8**	**9**	**10**
Sufficient	0.92099	0.91957	0.91786	0.91586	0.91358	0.91104	0.90822	0.90516	0.90184	0.89828	0.89448
SLS-failure	0.07643	0.07659	0.07705	0.07781	0.07885	0.08016	0.08173	0.08355	0.08561	0.08790	0.09041
ULS-failure	0.00258	0.00384	0.00509	0.00633	0.00757	0.00881	0.01005	0.01130	0.01255	0.01383	0.01511

Cont...

No_TD						**5**					
No_PD	**0**	**1**	**2**	**3**	**4**	**5**	**6**	**7**	**8**	**9**	**10**
Sufficient	0.90204	0.90033	0.89833	0.89605	0.89350	0.89069	0.88762	0.88430	0.88074	0.87694	0.87292
SLS-failure	0.09396	0.09412	0.09457	0.09532	0.09635	0.09764	0.09920	0.10099	0.10303	0.10529	0.10777
ULS-failure	0.00400	0.00555	0.00710	0.00863	0.01015	0.01167	0.01318	0.01470	0.01623	0.01777	0.01932

Cont...

No_TD						**6**					
No_PD	**0**	**1**	**2**	**3**	**4**	**5**	**6**	**7**	**8**	**9**	**10**
Sufficient	0.88340	0.88141	0.87913	0.87658	0.87377	0.87070	0.86738	0.86381	0.86002	0.85599	0.85174
SLS-failure	0.11089	0.11104	0.11149	0.11223	0.11324	0.11452	0.11606	0.11783	0.11984	0.12208	0.12452
ULS-failure	0.00571	0.00755	0.00938	0.01119	0.01299	0.01478	0.01657	0.01835	0.02014	0.02194	0.02374

Cont...

No_TD						**7**					
No_PD	**0**	**1**	**2**	**3**	**4**	**5**	**6**	**7**	**8**	**9**	**10**
Sufficient	0.86508	0.86280	0.86025	0.85744	0.85437	0.85105	0.84749	0.84369	0.83966	0.83541	0.83094
SLS-failure	0.12722	0.12737	0.12781	0.12854	0.12955	0.13081	0.13232	0.13408	0.13606	0.13827	0.14068
ULS-failure	0.00771	0.00983	0.01193	0.01402	0.01609	0.01814	0.02019	0.02224	0.02428	0.02633	0.02838

Bayes' theorem. Influence diagrams, which can be seen as an extension of Bayesian networks, are presented more in detail in *Section B.5*.

The testing model

The testing model built here consists of one decision node and six chance nodes, some of which are the same as for the main decision model, and are depicted in Figure B.7. Table B.9 gives a summary of the nodes in the testing model. The influence diagram is built and solved with the software Hugin Expert (Jensen et al 2002).

The node *QR*, or 'quality of old records', can be either 'good' or 'bad'. That is, with a good quality of old records we are very likely to be able to point out those piles that we can be suspicious of and hence the number of unknown defect piles will be less. With a bad quality of old records, we are less likely to locate all defect piles in the population.

The node *ES*, or the 'expected status of pile population', is a function of:

(1) the care taken when forming the initial piles,
(2) any corrosion or loss in durability that has taken place over the life of the piles, and
(3) the care taken when the first superstructure was demolished.

In the idealised case study, the expected status of pile population is in either one of three states: good, average or bad. The number of (unknown) defect piles is depending on the expected status of pile population, and we are likely to have less defect piles if the expected status of pile population is good rather than bad.

No_TD and *No_PD* are the prior estimates of the probabilities of there existing a certain number of total and partially defective piles, respectively, in the pile population. These estimated probabilities are based on our (to some degree subjective) knowledge about the overall quality of the whole pile population and are affected by the chance nodes *ES* and *QR*. These prior estimates are given in Table B.5, and again here in numbers in Tables B.10 and B.11. Notice that the probabilities presented in Table B.5 are estimated absolute maximum number of defective piles that are expected to be found in the population, whereas Tables B.10 and B.11 present an extended computation of probability of a specific number of defective piles to occur in the population based on exponential decay function with its maximum at 0 defects. From Table B.10 it can be seen that given that the expected status of the pile population is in state *bad* and the quality of old records is in state *bad*, then there is a probability of 0.44 (1.0–0.56) that there is at least one partially defective pile in the population. Furthermore, there is a probability of 0.1023 that there are exactly two partially defective piles in the pile population.

From the numbers in Table B.11 it can be seen that given that the expected status of the pile population is in state *good* and the quality of old records is in state *good*, the probability of there being defect piles is considerably lower. Given this prior information, there is a probability of 0.01 that there is at least one partially defective pile in the population and a probability of 0.00232 that there are exactly two partially defective piles in the pile population. The probability for any given number of defective piles is decreasing with an exponential function down to the maximum number of possible defect piles.

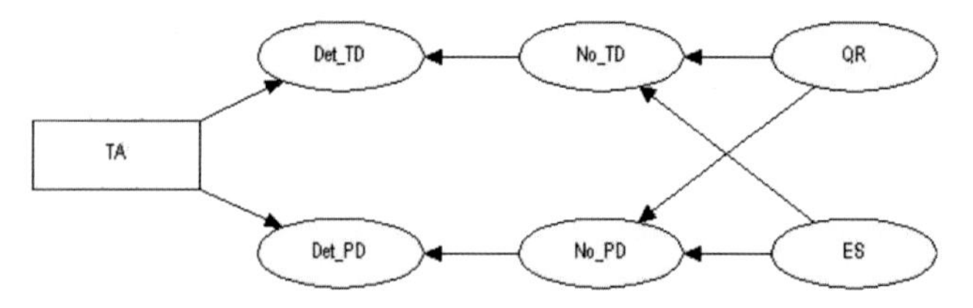

Figure B.7 The testing model built to provide input data to the main decision model

Table B.9 Summary of nodes in the testing model. The type of node, the name of the node, the short name of the node (or the symbol), and the number of states of each node are given

Type of node	Name of node	Symbol	Number of states of the node
Decision node	Testing alternatives	TA	2 (see below *)
Chance node	Number of detected totally defective piles in the test	Det_TD	8 (detecting 0–7 totally defective piles)
	Number of detected partially defective piles in the test	Det_PD	11 (detecting 0–10 partially defective piles)
	Number of totally defective piles in the pile population	No_TD	8 (0–7 totally defective piles)
	Number of partially defective piles in the pile population	No_PD	11 (0–10 partially defective piles)
	Quality of the old records	QR	2 (good, bad)
	Expected status of the pile population	ES	3 (good, average, bad)

* The decision node TA contains the three testing options:
Limited investigation taken as 3% of the pile population tested
Reasonable investigation taken as 10% of the pile population tested

Table B.10 Estimates of the prior probabilities of the number of partially defective piles in the pile population

QR	Good			Bad		
ES	**Good**	**Average**	**Bad**	**Good**	**Average**	**Bad**
No_PD ↓						
0	0,99	0,9	0,7	0,792	0,72	0,56
1	0,00632382	0,0632381	0,189714	0,131535	0,177067	0,278247
2	0,00232493	0,0232493	0,069748	0,0483585	0,065098	0,102297
3	0,000854754	0,00854754	0,0256426	0,0177789	0,0239331	0,0376091
4	0,000314248	0,00314248	0,00942744	0,00653636	0,00879894	0,0138269
5	0,000115532	0,00115532	0,00346597	0,00240307	0,0032349	0,00508342
6	0,000042475	0,000424751	0,00127425	0,000883482	0,0011893	0,0018689
7	0,000015616	0,000156159	0,000468476	0,000324811	0,000437245	0,000687098
8	0,000005741	0,000057411	0,000172234	0,000119415	0,000160751	0,00025261
9	0,000002111	0,000021107	0,000063321	0,000043903	0,000059100	0,000092871
10	0,000000776	0,000007760	0,000023280	0,000016141	0,000021728	0,000034144

Table B.11 Estimates of the prior probabilities of the number of totally defective piles in the pile population

QR	Good			Bad		
ES	**Good**	**Average**	**Bad**	**Good**	**Average**	**Bad**
No_PD ↓						
0	0,998	0,98	0,9	0,8982	0,882	0,81
1	0,00126586	0,0126586	0,0632928	0,0644323	0,0746857	0,120256
2	0,000465388	0,00465388	0,0232694	0,0236882	0,0274579	0,0442119
3	0,000171098	0,00171098	0,00855492	0,00870889	0,0100948	0,0162543
4	0,000062904	0,000629039	0,00314519	0,00320181	0,00371133	0,00597586
5	0,000023126	0,000231264	0,00115632	0,00117713	0,00136446	0,00219701
6	0,000008502	0,000085024	0,000425118	0,00043277	0,000501639	0,000807724
7	0,000003126	0,000031259	0,000156293	0,000159107	0,000184426	0,000296957

The chance nodes *Det_TD* and *Det_PD* contains the probabilities of detecting a certain number of defect piles, conditioned on how many defective piles there might be in the population (nodes *No_TD* and *No_PD*) and on the chosen investigation alternative. They are the link between the information from the testing procedure and the estimate of how many piles in the population that actually are defective. The probabilities in these nodes are calculated using a hypergeometric probability function (Levine et al 2001). In general, a hypergeometric distribution is used to determine the probability of x successes in n trials given: N = the population size and A = the number of successes in the population: $P(X = x|n, N, A)$. In the TM, it was assumed that the maximum number of totally defective piles possibly expected in the populations of 245 piles was 7, and that the maximum number partially defective piles possibly expected was 10.

For example, when using a hypergeometric density function for the reasonable investigation alternative (ie the number of piles being investigated, $n = 25$) and given that there are $A = 4$ totally defective piles out of $N = 245$, the probability of detecting exactly $x = 3$ totally defective piles in $n = 25$ trials equals 0.00345.

All calculated probabilities are given in Tables B.12–B.15. The chance nodes Det_TD and Det_PD are conditioned on the chance nodes No_TD and No_PD where the prior estimation of the probability of an existence of a certain number of total and partially defective piles in the population is stored. The chance nodes No_TD and No_PD are represented by A in the equation for the hypergeometric density function above.

Bayesian updating in the model

During the Bayesian updating process in the influence diagram, evidence (or observations) is introduced in the nodes Det_TD and Det_PD. Assuming that a reasonable investigation (ie $n = 25$) resulted in detecting two totally defective piles, this information is then introduced in node Det_TD as evidence (or an observation) and the node becomes instantiated. When the node Det_TD has been instantiated, new probabilities are calculated in node No_TD, using Bayes' theorem together with the prior information in the node No_TD and the information from the observation made.

The node No_TD is being updated by this procedure, such that the updated information in node No_TD now expresses the posterior probabilities of the existence of a specific number of defective piles. Assuming, for example, that QR is in state *bad*, ES is in state average then, when detecting two totally defective piles (ie observing 2 in Det_TD), the posterior probability of there being exactly two totally defective piles in the population is 0.303. Furthermore, the posterior probability of there being exactly three totally defective piles is 0.3026, and the posterior probability of

Table B.12 P(Det_TD/No_TD, TA = 3%) calculated from a hypergeometric density function

TA	3% (n=7 piles investigated with NDT)							
No_TD	0	1	2	3	4	5	6	7
Det_TD ↓								
0	1	0.967347	0.935631	0.904828	0.874916	0.845873	0.817678	0.790308
1	0	0.032653	0.063433	0.092408	0.119647	0.145214	0.169175	0.19159
2	0	0	0.000937	0.002741	0.005346	0.008688	0.012706	0.017342
3	0	0	0	0.000023	0.000091	0.000222	0.000434	0.000744
4	0	0	0	0	0	0.000002	0.000007	0.000016
5	0	0	0	0	0	0	0	0
6	0	0	0	0	0	0	0	0
7	0	0	0	0	0	0	0	0

Table B.13 P(Det_TD/No_TD, TA = 10%) calculated from a hypergeometric density function

TA	10% (n=25 piles investigated with NDT)							
No_TD	0	1	2	3	4	5	6	7
Det_TD ↓								
0	1	0.897959	0.805955	0.723038	0.648344	0.581088	0.520558	0.466107
1	0	0.102041	0.184008	0.248752	0.298776	0.336278	0.36318	0.381162
2	0	0	0.010037	0.02726	0.049339	0.074384	0.100883	0.127645
3	0	0	0	0.00095	0.003454	0.007848	0.014257	0.022653
4	0	0	0	0	0.000086	0.000394	0.001079	0.002297
5	0	0	0	0	0	0.000008	0.000041	0.000133
6	0	0	0	0	0	0	0.000001	0.000004
7	0	0	0	0	0	0	0	0

Table B.14 P(Det_PD/No_PD, TA = 3%) calculated from a hypergeometric density function

TA	3% (n=7 piles investigated with NDT)										
No_PD	0	1	2	3	4	5	6	7	8	9	10
Det_PD ↓											
0	1	0.96735	0.93563	0.90483	0.87492	0.84587	0.81768	0.79031	0.76374	0.73796	0.71295
1	0	0.03265	0.06343	0.09241	0.11965	0.14521	0.16918	0.19159	0.21252	0.23202	0.25016
2	0	0	0.00094	0.00274	0.00535	0.00869	0.01271	0.01734	0.02254	0.02825	0.03441
3	0	0	0	0.00002	0.00009	0.00022	0.00043	0.00074	0.00117	0.00171	0.00239
4	0	0	0	0	0	0	0.00001	0.00002	0.00003	0.00006	0.00009
5	0	0	0	0	0	0	0	0	0	0	0
6	0	0	0	0	0	0	0	0	0	0	0
7	0	0	0	0	0	0	0	0	0	0	0
8	0	0	0	0	0	0	0	0	0	0	0
9	0	0	0	0	0	0	0	0	0	0	0
10	0	0	0	0	0	0	0	0	0	0	0

Table B.15 P(Det_PD/No_PD, TA = 10%) calculated from a hypergeometric density function

TA	10% (n=25 piles investigated with NDT)										
No_PD	0	1	2	3	4	5	6	7	8	9	10
Det_PD ↓											
0	1	0.89796	0.80596	0.72304	0.64834	0.58109	0.52056	0.46611	0.41715	0.37314	0.33362
1	0	0.10204	0.18401	0.24875	0.29878	0.33628	0.36318	0.38116	0.39169	0.39603	0.39528
2	0	0	0.01004	0.02726	0.04934	0.07438	0.10088	0.12765	0.15375	0.17849	0.20137
3	0	0	0	0.00095	0.00345	0.00785	0.01426	0.02265	0.03289	0.04476	0.05798
4	0	0	0	0	0.00009	0.00039	0.00108	0.0023	0.00419	0.00687	0.01043
5	0	0	0	0	0	0.00001	0.00004	0.00013	0.00032	0.00067	0.00122
6	0	0	0	0	0	0	0	0	0.00002	0.00004	0.00009
7	0	0	0	0	0	0	0	0	0	0	0.00001
8	0	0	0	0	0	0	0	0	0	0	0
9	0	0	0	0	0	0	0	0	0	0	0
10	0	0	0	0	0	0	0	0	0	0	0

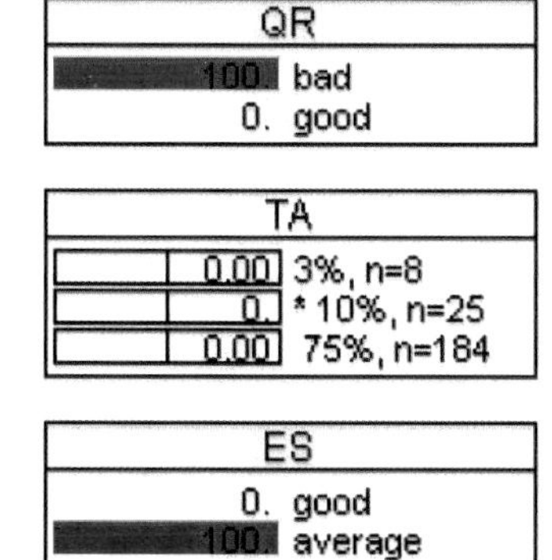

Figure B.8 The result from the testing model when QR is bad, ES is average, the testing alternative is to test 10% of the pile population and the observation made was two detected totally defective piles

there being exactly four totally defective piles is 0.2013 (Figure B.9).

In the testing model, used for producing input data to the main decision model, it was assumed that zero total and zero partially defective piles were detected in both investigation programs. The posterior probabilities in *No_PD* and *No_TD* were used as input data. All input data probabilities for the chance nodes *No_TD* and *No_PD* are given in Tables B.16–B.19 for each decision alternative.

B.5 Bayesian networks and influence diagrams: theory

A Bayesian network (BN) consists of a directed acyclic graph that describes dependancies between probabilistic variables and influence diagrams as an extension of Bayesian networks. The following introduction to Bayesian networks and influence diagrams is based on Jensen (2001).

Consider the following joint probability table for variables *A* and *B*:

$P[B, A]$	a_1	a_2
b_1	0.12	0.48
b_2	0.28	0.12

From table $P[B, A]$, the probability distribution $P[A]$ can be calculated by variable B being marginalized out of $P[B, A]$: $P[A] = [0.4, 0.6]$. In the same way $P[B] = [0.6, 0.4]$. The fundamental rule, $P[B|A]\,P[A] = P[A, B]$, makes it possible to calculate both $P[B|A]$ and $P[A|B]$. If evidence (e) is received that $A = a_1$, then $P[B, a_1] = [0.12, 0.28]$. Calculating $P[B|a_1]$, using the fundamental rule yields:

$$P[B|a_1] = \frac{P[B, a_1]}{P[a_1]} = [0.3, 0.7]$$

Thus, by having access to the joint probability table for a set of uncertain variables, it is possible to perform reasoning under uncertainty. However, joint probability tables grow exponentially with the number of variables and a Bayesian network is a compact way of representing large joint probability tables. The tables are generally called potential tables. The definition of Bayesian networks (BN) is that they consist of the following.

- A set of variables and a set of directed edges between variables.
- Each variable has a finite set of mutually exclusive states.[1]
- The variables together with the directed edges form a directed acyclic graph (DAG). A directed graph is acyclic if there is no directed path $A_1 \rightarrow \cdots \rightarrow A_n$ such that $A_1 = A_n$.
- To each variable A with parents $B_1, \ldots, B_n$, the potential table $P[A| B_1, \ldots, B_n]$ is attached. If A has no parents, then the table is reduced to unconditional probabilities $P[A]$.

Let $U = \{A_1, \ldots, A_n\}$ be a universe of variables. A BN over U is a representation of the joint probability table $P[U]$ and can be calculated from the potentials specified in the network. The chain rule for BNs is as follows: let BN be a Bayesian network over $U = \{A_1, \ldots, A_n\}$. Then, the joint probability distribution $P[U]$ is the product of all potentials specified in BN:

$$P[U] = \prod_i P[A_i|pa(A_i)]$$

where $pa(A_i)$ is the parent set of A_1. Reasoning under uncertainty and calculations by introducing evidence can be performed by a method called bucket elimination (for full information, the reader is referred to Jensen (2001) without having to deal with the full joint probability table. A *simple* BN with two variables is shown in Figure B.9.

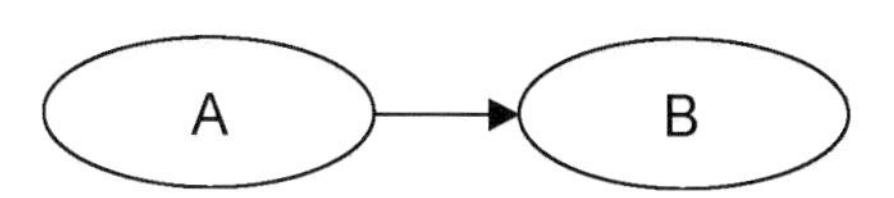

Figure B.9 A simple Bayesian network with two variables, A and B. B is a child of A, and A is a parent of B. B is conditionally dependent on A

[1] Mutually exclusive means that the variable of the node can only be in one of the states, and must be in one of the states.

Table B.16 Probabilities for totally defective piles when the quality of old records (QR) is GOOD

ES	GOOD		
RA	Complete replacement	Full re-use reasonable investigation	Full re-use limited investigation
No_TD ↓			
0 TD	0.99998	0.99830390	0.99810102
1 TD	0.00002	0.00113704	0.00122465
2 TD	0	0.00037520	0.00043548
3 TD	0	0.00012375	0.00015483
4 TD	0	0.00004080	0.00005504
5 TD	0	0.00001344	0.00001956
6 TD	0	0.00000443	0.00000695
7 TD	0	0.00000146	0.00000247
ES	AVERAGE		
RA	Complete replacement	Full re-use reasonable investigation	Full re-use limited investigation
No_TD ↓			
0	0.99998	0.98299235	0.98099285
1	0.00002	0.01140161	0.01225766
2	0	0.00376227	0.00435872
3	0	0.00124088	0.00154971
4	0	0.00040908	0.00055091
5	0	0.00013480	0.00019582
6	0	0.00004439	0.00006959
7	0	0.00001461	0.00002473
ES	BAD		
RA	Complete replacement	Full re-use reasonable investigation	Full re-use limited investigation
No_TD ↓			
0	0.99998	0.91391045	0.90457772
1	0.00002	0.05771277	0.06153752
2	0	0.01904395	0.02188229
3	0	0.00628114	0.00778010
4	0	0.00207068	0.00276577
5	0	0.00068231	0.00098308
6	0	0.00022472	0.00034938
7	0	0.00007398	0.00012415

A *serial connection* is shown in Figure B.10. *A* has an influence on *B*, which in turn has an influence on *C*. Evidence (*e*) on *A* will influence the certainty of *B*, which then influences the certainty on *C*. Similarly, evidence on *C* will influence the certainty on *A* through *B*. If the state of *B* is known, then the channel is blocked, and *A* and *C* become independent: *A* and *C* are d-separated given *B*. When the state of a variable is known, it is instantiated. Evidence may be transmitted through a serial connection unless the state of the variable in the connection is known.[2]

Figure B.11 shows a *diverging* connection. Influence can pass between all the children of *A* unless the state of *A* is known: *B*, *C* and *E* are d-separated given *A*. Thus, evidence may be transmitted through a diverging connection unless it is instantiated.

2 Evidence on a variable is a statement of the certainties of its states. If the variable is instantiated, it is called hard evidence; otherwise, it is termed soft evidence.

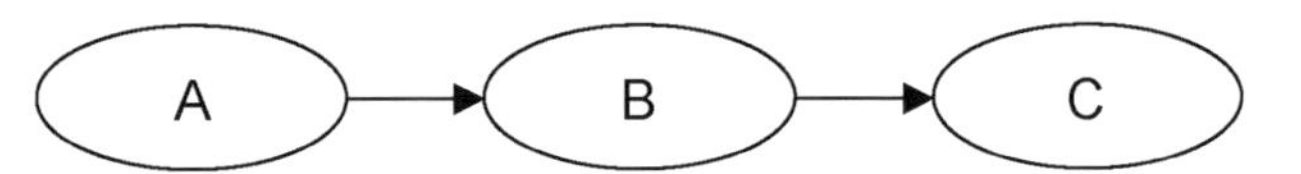

Figure B.10 Serial connection. When B is instantiated, it blocks communication between A and C. A and C are conditionally independent given B

A *converging* connection is shown in Figure B.12. If nothing is known about *A* except what may be inferred from knowledge of its parents *B*,..., *E*, then the parents are independent: evidence on them has no influence on the certainty of the others. Knowledge of one possible cause of an event does not tell us anything about other possible causes. However, if something is known about the consequences, information on one possible cause can yield information about the other causes. This is the *explaining-away* effect: *a*

Table B.17 Probabilities for totally defective piles when the quality of old records (QR) is BAD

ES	GOOD		
RA	Complete replacement	Full re-use reasonable investigation	Full re-use limited investigation
No_TD ↓			
0 TD	0.99998	0.91233630	0.90285107
1 TD	0.00002	0.05876815	0.06265114
2 TD	0	0.01939210	0.02227816
3 TD	0	0.00639596	0.00792085
4 TD	0	0.00210855	0.00281582
5 TD	0	0.00069478	0.00100086
6 TD	0	0.00022883	0.00035570
7 TD	0	0.00007533	0.00012639
ES	AVERAGE		
RA	Complete replacement	Full re-use reasonable investigation	Full re-use limited investigation
No_TD ↓			
0	0.99998	0.89813075	0.88729834
1	0.00002	0.06829123	0.07268099
2	0	0.02253456	0.02584476
3	0	0.00743241	0.00918893
4	0	0.00245023	0.00326661
5	0	0.00080737	0.00116109
6	0	0.00026591	0.00041264
7	0	0.00008753	0.00014663
ES	BAD		
RA	Complete replacement	Full re-use reasonable investigation	Full re-use limited investigation
No_TD ↓			
0	0.99998	0.83412267	0.81786415
1	0.00002	0.11120087	0.11745870
2	0	0.03669398	0.04176760
3	0	0.01210248	0.01485014
4	0	0.00398980	0.00527914
5	0	0.00131468	0.00187644
6	0	0.00043299	0.00066687
7	0	0.00014254	0.00023697

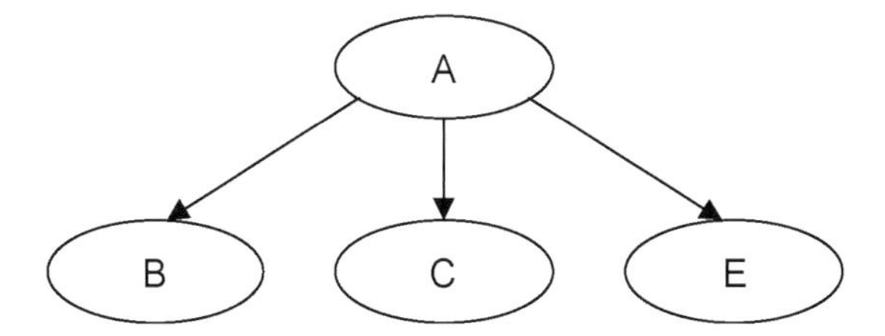

Figure B.11 Diverging connection. If A is instantiated, it blocks for communication between its children. This means that B, C and E are conditionally independent given A

Figure B.12 Converging connection. If A changes certainty, it opens for communication between its parents.

has occurred, and b as well as c may cause a. If there is information that c has occurred, the certainty of b will decrease. If there is information that c has not occurred, the certainty of b will increase. Evidence may only be transmitted through a converging connection if either the variable in the connection or one of its descendants has received evidence.

The three preceding cases cover all ways in which evidence may be transmitted through a variable, and it is thus possible to decide for any pair of variables in a causal

Table B.18 Probabilities for partially defective piles when the quality of old records (QR) is GOOD

ES	GOOD		
RA	Complete replacement	Full re-use reasonable investigation	Full re-use limited investigation
No_PD ↓			
0 PD	0.9998	0.99151300	0.99050294
1 PD	0.0002	0.00568721	0.00612044
2 PD	0	0.00187665	0.00217638
3 PD	0	0.00061896	0.00077380
4 PD	0	0.00020405	0.00027508
5 PD	0	0.00006724	0.00009778
6 PD	0	0.00002214	0.00003475
7 PD	0	0.00000729	0.00001235
8 PD	0	0.00000240	0.00000439
9 PD	0	0.00000079	0.00000156
10 PD	0	0.00000026	0.00000055
ES	AVERAGE		
RA	Complete replacement	Full re-use reasonable investigation	Full re-use limited investigation
No_PD ↓			
0 PD	0.9998	0.91394649	0.90459323
1 PD	0.0002	0.05766517	0.06148539
2 PD	0	0.01902825	0.02186376
3 PD	0	0.00627597	0.00777352
4 PD	0	0.00206898	0.00276344
5 PD	0	0.00068175	0.00098224
6 PD	0	0.00022453	0.00034908
7 PD	0	0.00007391	0.00012404
8 PD	0	0.00002432	0.00004407
9 PD	0	0.00000800	0.00001566
10 PD	0	0.00000263	0.00000556
ES	BAD		
RA	Complete replacement	Full re-use reasonable investigation	Full re-use limited investigation
No_PD ↓			
0 PD	0.9998	0.73358276	0.71082821
1 PD	0.0002	0.17852826	0.18635811
2 PD	0	0.05891062	0.06626780
3 PD	0	0.01943007	0.02356105
4 PD	0	0.00640547	0.00837581
5 PD	0	0.00211066	0.00297712
6 PD	0	0.00069514	0.00105804
7 PD	0	0.00022884	0.00037597
8 PD	0	0.00007529	0.00013358
9 PD	0	0.00002476	0.00004745
10 PD	0	0.00000814	0.00001685

network whether they are independent given the evidence entered in the network. The definition of BNs does not refer to causality, and there is no requirement that the links represent causal impact. Instead, it is required that *the d-separation properties implied by the structure hold*. There is, however, a good reason to strive for causal networks since a correct model of a causal domain is minimal with respect to links.

According to (McCabe et al 1998), BN applications, including diagnostics, forecasting and decision support, have been used in, among others, the medical and software development fields. There are some examples of studies using BNs in civil engineering and environmental risk management applications (Chong & Walley 1996, McCabe et al 1998, Pendock & Sears 2002, Sahely & Bagley 2001, Varis 1998, Varis & Kuikka 1997). Several free-ware and

Table B.19 Probabilities for partially defective piles when the quality of old records (QR) is BAD

ES	GOOD		
RA	Complete replacement	Full re-use reasonable investigation	Full re-use limited investigation
No_PD ↓			
0 PD	0.9998	0.81796235	0.80045426
1 PD	0.0002	0.12198487	0.12859822
2 PD	0	0.04025240	0.04572864
3 PD	0	0.01322762	0.01625857
4 PD	0	0.00437673	0.00577981
5 PD	0	0.00144217	0.00205439
6 PD	0	0.00047498	0.00073012
7 PD	0	0.00015636	0.00025944
8 PD	0	0.00005145	0.00009218
9 PD	0	0.00001692	0.00003274
10 PD	0	0.00000556	0.00001163
ES	AVERAGE		
RA	Complete replacement	Full re-use reasonable investigation	Full re-use limited investigation
No_PD ↓			
0 PD	0.9998	0.75213632	0.73038404
1 PD	0.0002	0.16609563	0.17375555
2 PD	0	0.05480782	0.06178607
3 PD	0	0.01807691	0.02196766
4 PD	0	0.00595937	0.00780936
5 PD	0	0.00196366	0.00277578
6 PD	0	0.00064673	0.00098649
7 PD	0	0.00021290	0.00035054
8 PD	0	0.00007005	0.00012454
9 PD	0	0.00002304	0.00004424
10 PD	0	0.00000757	0.00001571
ES	BAD		
RA	Complete replacement	Full re-use reasonable investigation	Full re-use limited investigation
No_PD ↓			
0 PD	0.9998	0.60030634	0.57279767
1 PD	0.0002	0.26783782	0.27531254
2 PD	0	0.08838094	0.09789946
3 PD	0	0.02915003	0.03480745
4 PD	0	0.00960983	0.01237384
5 PD	0	0.00316652	0.00439820
6 PD	0	0.00104289	0.00156308
7 PD	0	0.00034331	0.00055543
8 PD	0	0.00011296	0.00019734
9 PD	0	0.00003715	0.00007010
10 PD	0	0.00001221	0.00002490

commercial software programs are available on the internet [eg Hugin Expert (Jensen et al 2002) and Genie (Decision Systems Laboratory 2003)].[3]

As BNs contain only chance nodes, an influence diagram (ID) consists of a directed acyclic graph over chance nodes (probabilistic variables), decision nodes and utility nodes (deterministic variables), with a directed path that includes all decision nodes. IDs were originally invented as a compact representation of decision trees for symmetric decision scenarios. Jensen (2001) argues that today, IDs are often considered as a decision tool extending BNs.

IDs consist of a directed acyclic graph over chance nodes, decision nodes and utility nodes with the following structural properties: a directed path comprising all decision nodes

3 A list of software packages for building graphical models can be found at: http//www.ai.mit.edu/~murphk/Bayes/bnsoft.html.

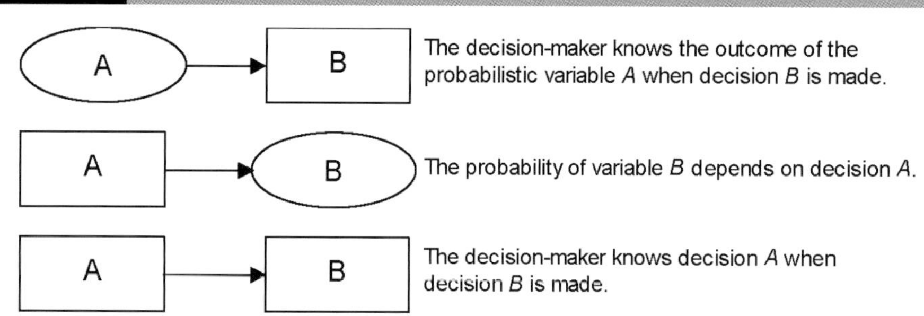

Figure B.13 Interpretation of links in influence diagrams. Modified from Attoh-Okine (1998)

(ie there is a temporal sequence of decisions) and the utility nodes have no children. Quantitative specifications require that:

(1) decision and chance nodes have a finite set of mutually exclusive states,

(2) the utility nodes have no states,

(3) a conditional probability table $P[A|pa(A)]$ is attached to each chance node A, and finally,

(4) a real-valued function over $pa(U)$ is attached to each utility node U.

Links pointing to decision nodes (information links) indicate that the state of the parent is known prior to the decision while links pointing to chance nodes indicate the conditioning of variables (Figure B.13). The assumption of no-forgetting is made here as well, ie the decision maker remembers the past observations and decisions. In common with decision trees, a decision strategy is desired when the ID is solved and the basic requirement is to obtain a recommendation for the first decision. IDs are solved according to the principle of expected utility. Jensen (2001) provides some algorithms for IDs and Henrion (1989) presents some practical issues in constructing BN and IDs.

Software programs for BNs are more common than those for IDs, although there are several ID programs available [eg Hugin (Jensen et al 2002) and Genie (Decision Systems Laboratory 2003)].[4] Other ID applications found in the literature are, eg Hong & Apostolakis (1993) and (Jeljeli & Russell 1995). Two interesting papers by Varis (1997) and Varis & Kuikka (1999) summarise the experience gained through application of BNs and IDs.

The probabilities used as input in Bayesian networks or IDs can be estimated by expert judgements or be calculated by probabilistic models. An advantage of using IDs or BNs is the possibility of combining both objectively estimated probabilities and subjectively estimated probabilities, thus making use of all available knowledge.

4 A list of software packages for building graphical models can be found at: http//www.ai.mit.edu/~murphk/Bayes/bnsoft.html.

B.6 References

Attoh-Okine NO. Potential application of influence diagram as a risk assessment tool in brownfields sites. In: Hoddinott KB (ed) *Superfund risk assessment in soil contamination studies.* 3rd Volume, ASTM STP **1338:** 148–159. American Society for Testing and Materials.1998

British Standards Institution. BS EN 1990: 2002 *Eurocode. Basis of structural design*

Cameron G & Chapman T. Quality assurance of bored pile foundations. *Ground Engineering* 2004: **37**(2): 35–40

Chapman TJP. Foundations through congested urban ground. *Proceedings of XIII European Conference on Soil Mechanics and Geotechnical Engineering Design,* Prague, 2003, pp 383–386

Chapman T & Marcetteau A. Achieving economy and reliability in piled foundation design for a building project. *The Structural Engineer* 2004: **82** (11)

Chong HG & Walley WJ. Rule-based versus probabilistic approaches to the diagnosis of faults in wastewater treatment processes. *Artificial Intelligence in Engineering* 1996: **10**(3): 265–273

Davis AG & Dunn CS. From theory to field experience with the non-destructive vibration testing of piles. *Proceedings of Institution of Civil Engineers* 1974: **57**(2): 571–593

Decision Systems Laboratory. *Genie 2.* 2004-06-29. Decision Systems Laboratory, University of Pittsburgh. 2003. http://www.sis.pitt.edu/~genie/.

Ellingwood BR. Probability-based structural design: prospects for acceptable risk bases. In: Melchers RE & Stewart MG (eds) *Applications of statistics and probability.* pp 11–18. Balkema, Rotterdam, 2000

Ellway K. Practical guidance on the use of intégrity tests for quality control of cast-in-situ piles. *Proceedings of the 1st International Conference on Foundations and Tunnels,* London, 1987. pp 228–234

Fleming WGK, Weltman AJ, Randolph MF & Elson WK. *Piling engineering.* Guildford, Surrey University Press, 1985

Henrion M. Some practical issues in constructing belief networks. In: Kanal LN, Levitt TS & Lemmer JF (eds) *Uncertainty in artificial intelligence 3.* Vol 8, pp 161–173. Amsterdam, Elsevier. 1989

Hong Y & Apostolakis G. Conditional influence diagrams in risk management. *Risk Analysis* 1993: **13**(6): 625–636

Jeljeli MN & Russell JS. Coping with uncertainty in environmental construction: Decision-analysis approach. *Journal of Construction Engineering and Management* 1995: **121**(4): 370–380

Jensen FV. *Bayesian networks and decision graphs.* New York, Springer, 2001

Jensen F, Kjærulff UB, Lang M & Madsen AL. HUGIN: the tool for Bayesian networks and influence diagrams. In: Gámez JA & Salmerón A (eds) *Proceedings of the 1st European Workshop on Probabilistic Graphical Models,* Cuenca, Spain, 2002. pp 212–221

Levine DM, R PP & Smidt RK. Applied statistics for engineers and scientists. *Using Microsoft Excel and MINITAB.* Prentice Hall, Upper Saddle River, New Jersey, 2001

Lew M, Zadoorian CJ & Carpenter LD. Integrity testing of drilled piles for tall buildings. *Structure Magazine*, jointly published by National Council Probabilistic Methods in Geotechnical Engineering and US National Academy Press, 2002

McCabe B, AbouRizk SM & Goebel R. Belief networks for construction performance designs. *Journal of Computing in Civil Engineering* 1998: **12**(2): 93–100

Pendock NE & Sears M. Choosing geological models with Bayesian belief networks. *South African Journal of Science* 2002: **98** (September/October): 500–502

Preiss K & Shapiro J. Statistical considerations in pile testing. Congress on the Mechanics of Soils, Stockholm, 1981. pp 799–802

Rice JA. *Mathematical statistics and data analysis.* 2nd edition. Belmont CA, Duxbury Press, 1995

Sahely BSGE & Bagley DM. Diagnosing upsets in anaerobic wastewater treatment using Bayesian belief networks. *Journal of Environmental Engineering* 2001: **127**(4): 302–310

Thasnanipan N, Maung AW & Baskaran G. Sonic integrity test on piles founded in Bangkok subsoil: signal characteristics and their interpretations. *Proceedings of the 4th International Conference on Case Histories in Geotechnical Engineering*, St.Louis, Missouri, 9–12 March, 1988, pp 1086–1092

Varis O. Bayesian decision analysis for environmental and resource management. *Environmental Modelling & Software* 1997: **12**(2–3): 177–185

Varis O. Belief network approach to optimization and parameter estimation: Application to resource and environmental management. *Artificial Intelligence* 1998: **101:** 135–163

Varis O & Kuikka S. BeNe-EIA: A Bayesian approach to expert judgement elicitation with case studies on climate change impacts on surface waters. *Climatic Change* 1997: **37:** 539–563

Varis O & Kuikka S. Learning Bayesian decision analysis by doing: lessons from environmental and natural resources management. *Ecological Modelling* 1999: **119:** 177–195

Vose D. Quantitative risk analysis. In: *A guide to Monte Carlo simulation modelling*. Chichester, Wiley, 1996

Appendix C: Whole life cost and environmental impact case studies

C.1 Whole life cost (WLC) example

A WLC analysis for an idealised redevelopment of an office building is given as a case study to illustrate the WLC process.

The options in the WLC model are:

- demolish and rebuild, with new foundations,
- demolish with intention of using old foundations and rebuild with same (compromised) layout (using either some or all of the old foundations),
- demolish with intention of using all old foundations and rebuild with same (compromised) layout.

Clearly, in some cases, do nothing will be another option!

The example is used to illustrate the WLC stages for each option, calculation of total costs and then calculation of costs discounted to present day. Generally, a number of different building lifespans would be considered.

Table C.1 shows the cost assumptions for each option, at each stage of redevelopment and life of the building. These are summarised in Table C.2, either undiscounted or discounted to NPV using 3.5% pa.

The costs are dominated by those incurred during the building life, although any savings made in construction do have an impact. For a greenfield site, design with reuse in mind has a lower WLC when future site options are

Table C.1 Costs at each stage

		Cost for each scenario (£k) -ve costs are income					
		Greenfield site		Brownfield site			
Stage	Activity	Greenfield	Greenfield design for re-use	No reuse	Partial reuse, no compromise on layout or form	Partial reuse, compromised layout	Full reuse, compromised layout
Purchase	Buy land	25000	25000	15000	15000	15000	15000
Demolition	Source old design/construction records and verify			5	5	5	5
	Install building monitoring/ monitor/ analyse/ report				20	20	40
	Carry out load takedown/ analyse/ report				5	5	5
	Demolish superstructure			1000	1100	1100	1100
	Remove old foundations/substructure			500		250	
Design	SI for normal foundation design process	40	40	40	40	40	40
	SI for re-use				20	20	20
	SI for avoidance			20			
	SI for removal			10			
	Make records	2	2	2	2	2	2
	Foundation design for new foundations	20	20	30	20	20	
	Foundation design for new foundations to replace removed			20			
	Foundation design for old foundations				15	15	35
	Foundation design for mixed scheme foundations				5	5	
	Records	5	5	5	7.5	7.5	7.5
	Specifications for construction/ monitoring	1	1	1.5	2	2	2
	Other substructures	10	15				
	Transfer structures					10	10
	Superstructure	500	600	700	500	700	700
Construction	Construction of new foundation elements	570	640	700	300	300	
	Other substructure elements	1800	2000	2500	1800	2200	2400
	Records	5	5	5	5	5	5
	Insurance	25	25	30	35	35	40
	Upgrading of old foundations				50	50	100
	Testing/ monitoring	20	20	20	30	30	40
	Superstructure	10000	12000	15000	12000	15000	15000
	Fit-out & finishes	17000	17000	17000	17000	17000	17000
In service	Normal in service energy, maintenance (25 years)	27000	27000	27000	27000	27000	27000
	Compromised layout - increased energy use, maintenance (25 years)					3000	3000
	Rent over 25 years	-83750	-83750	-83750	-83750		
	Compromised layout rent over 25 years					-82500	-82500
	Business costs (25)	781000	781000	781000	781000	760000	760000
	Insurance (25)	2500	2500	2500	2500	2500	2500
	Monitoring and keeping records		50		50	50	50
Future	Redevelopment costs	4000	3500	4500	4200	4200	4000
	Redevelopment options		-5000				-2500
	Residual building value	-10000	-10000	-10000	-10000	-9000	-9000
	Residual land value	-3000	-4000	-2500	-3000	-3000	-3500

Table C.2 Summary of costs

1. No discounting 0

Assumed time period Scenarios

Life stage	Start	End	Greenfield	Greenfield design for reuse	Partial reuse, no compromise on layout or form	Partial reuse, compromise on layout	Full reuse, compromise on layout
Purchase	0	0.1	25000	25000	15000	15000	15000
Demolition	0.1	0.5	0	0	1505	1130	1380
Investigation a	0	1	578	683	828.5	611.5	821.5
Construction	0.8	2.8	29420	31690	35255	31220	34620
In service	2	27	726750	726800	726750	726800	710050
Future	27	30	-9000	-15500	-8000	-8800	-7800
Totals			772748	768673	771339	765962	754072

2. Discount @ x% per year (in service and future only) 0.035 =x

Scenarios

Life stage	Start	End	Greenfield	Greenfield design for reuse	Partial reuse, no compromise on layout or form	Partial reuse, compromise on layout	Full reuse, compromise on layout
Purchase	0	0.1	25000	25000	15000	15000	15000
Demolition	0.1	0.5	0	0	1505	1130	1380
Investigation a	0	1	578	683	828.5	611.5	821.5
Construction	0.8	2.8	29420	31690	35255	31220	34620
In service	2	27	462915.6	462947.4	462915.6	462947.4	452278.2
Future	27	30	-3378.81	-5819.061	-3003.386	-3303.725	-2928.302
Totals			514535	514501	512501	507605	501171

considered, even though a small premium is paid at the time of construction. When redevelopment is considered, the WLC model implies that the reuse options offer a cost advantage over new build.

C.2 Environmental impact case study

This case study considers the environmental impact of a development solely by measuring the relative energy consumption of two options for foundation construction. This ignores all other environmental impacts made during and after construction (eg water use), but is seen as one way of comparing two options in simple terms.

- No foundation reuse, foundation slab and piles were broken out to 3 m depth and removed off site. New piles and a new slab were installed over the footprint, requiring a piling platform (Box C.1).
- Full reuse of the existing slab and piles where the old and new footprints coincided. Piles outside the new footprint broken out to 2 m to provide garden areas (Box C.2).

Box C.1 No foundation reuse: procedure

Raft broken out

- 0.9 m thick 43.3 m × 19.2 m concrete raft broken up and removed (area 831 m^2).
- 20 t excavator and dump truck working on site for 1 week.

Piles broken out to 3 m

- 351 piles of 0.45 m diameter broken out.
- 20 t excavator and dump truck working on site for 1 week.

Raft and piles removed off site

- Concrete removed in 12 t lorry loads to nearby (20 km) transfer station for crushing.

Platform imported over new building area

- New building area on old footprint covers 486 m^2.
- Crushed concrete imported in 12 t lorry loads from nearby (20 km) transfer station.
- 500 mm platform required over filled ground (old footprint), minimum depth elsewhere.

New piles and raft installed in new area

- 300 new piles, 0.5 m diameter 15 m length.
- 20 days piling rig, attendant machines on site.
- New raft 0.6 m deep.
- Readymix concrete brought from nearest depot (20 km) in 6 m^3 trucks.

Box C.2 Full reuse over coincident footprint: procedure

Raft broken out in garden areas only
- 0.9 m thick concrete raft broken out (area 345 m^2).
- 20 t excavator and dump truck working on site for 3 days.

Piles broken out to 2 m in garden areas
- 180 piles of 0.45 m diameter broken out.
- 20 t excavator and dump truck working on site for 3 days.

Raft and piles crushed
- Concrete crushed on site.

Raft and piles used on site to form new platform
- New building area on old footprint covers 486 m^2.
- All concrete reused on site to form minimum platform.

New piles installed in new area outside old footprint
- 150 new piles, 0.5 m diameter 15 m length.
- 10 days piling rig, attendant machines on site.
- New raft 0.9 m deep.
- Readymix concrete brought from nearest depot (20 km) in 6 m^3 trucks.

Basis of calculations

The time and machinery required on site was observed during demolition and site preparation. The requirements of the piling operation and likely timescales have been estimated based on previous experience and the piling specification. Where reuse of the piles could have taken place, these activities have been proportionately reduced.

Boxes C.3 and C.4 show the data used for the energy consumption for various activities.

Energy used during demolition and construction activities

During demolition of the existing raft, work on site took 2 weeks and an excavator and dumper were required. Percentage working hours were assumed. Given the time of year, setting out required to maintain the pile grid and steel reinforcement removal.

An excavator was required to construct the piling platform which had to be significantly deeper where piles had been removed and were to be replaced as the upper ground was disturbed.

During piling by a CFA rig, work took 3 weeks (average 20 piles per day during 5 full working hours, making allowance for setting out, downtime, concrete deliveries, etc.). Also required was an excavator. Use of the concrete pump has been neglected.

Energy used in transport

Transportation of the concrete off site has been assumed to be by a fixed wheel lorry capable of carrying 12 tonnes per load, and the distance to the local transfer station was 20 km.

Embodied energy in materials used

The concrete was considered to require 1% reinforcement using recycled steel bar.

Box C.3 Energy consumption by machinery

Transport and machinery outputs

● 12 t lorry cap	0.3576 litres/km	(DETR 1977, from BR 370*)
● 6 m^3 concrete	0.4155 litres/km	(DETR 1977, from BR 370*)
● Concrete crusher	8.35 MJ/tonne	(SimaPro LCA software)
● Excavator	100 kW 80% engine output	(Liebherr ~20 t)
● Piling rig	200 kW 50% engine output	(Liebherr ~20 tm)

* **Howard N, Edwards S & Anderson J.** *BRE methodology for environmental profiles of construction materials, components and buildings.* BR 370. Watford, IHS BRE Press, 1999

Box C.4 Embodied energy (MJ/kg). Data from BR 370*

● Concrete	2 MJ/kg
● Steel, recycled	10 MJ/kg
● Diesel	42 MJ/kg+

* **Howard N, Edwards S & Anderson J.** *BRE methodology for environmental profiles of construction materials, components and buildings.* BR 370. Watford, IHS BRE Press, 1999

\+ equivalent to 35 MJ/l

Results

Table C.3 shows the relative magnitudes of energy consumption at each stage.

Conclusions

- The energy use in constructing the new foundations could have been reduced by 50% through reusing the foundations where the old and new building footprints coincided.
- The original building had been significantly higher than the new construction, and hence the reuse of the old piles should have been feasible, although the calculations assume that the existing raft thickness is maintained over the new building.
- The data indicate that significant savings in transport energy could have been made by reusing the demolition materials on site.

Table C.3 Energy consumption at each stage

	Vol m^3	Mass concrete Mg	Energy in demolition/ construction	MJ	Embodied energy	MJ	Transport	MJ	TOTAL MJ
No reuse									
Raft broken out	748		1wk machines, 20t excavator +dump truck	18000					
Piles broken out to 3m	167		1wk machines, 20t excavator +dump truck	18000					
Raft and piles removed off site	916	21977	concrete crushed	183506			agg lorries 12t ea, 20 km	457386	
Platform imported over new building area	342	6156	1 day machine	1800			agg lorries 12t ea, 20 km	128121	
New piles and raft (0.6m) installed in new area	1424	34166	3 weeks piling rig	81000	concrete and reinforcement	71748	concrete lorries 6m3 ea 20km, spoil off site, 100km	4318906	
									5,278,467
Reuse of piles where footprints coincide									
Raft broken out in garden areas only	301		3 days 2 machines	10800					
Piles broken out to 2m in garden areas	45		2 days 2 machines	7200					
Raft and piles crushed	345	8290	concrete crusher	73698					
Raft and piles used on site to form new platform	333		1 day machine	1800					
New piles installed in new area outside old footprint	177	4259	1-2 weeks piling rig	40500	concrete and reinforcement	41045	concrete lorries 6m3 ea 20km, spoil off site, 100km	2470726	
									2,645,769

Appendix D: Flow charts and guidelines for integrity testing of foundations

D.1 Introduction

The following flow charts provide robust protocols that can be used to classify piles and other foundation elements for reuse using NDT methods.

Tables D1–D5 give general guidelines for NDT testing.

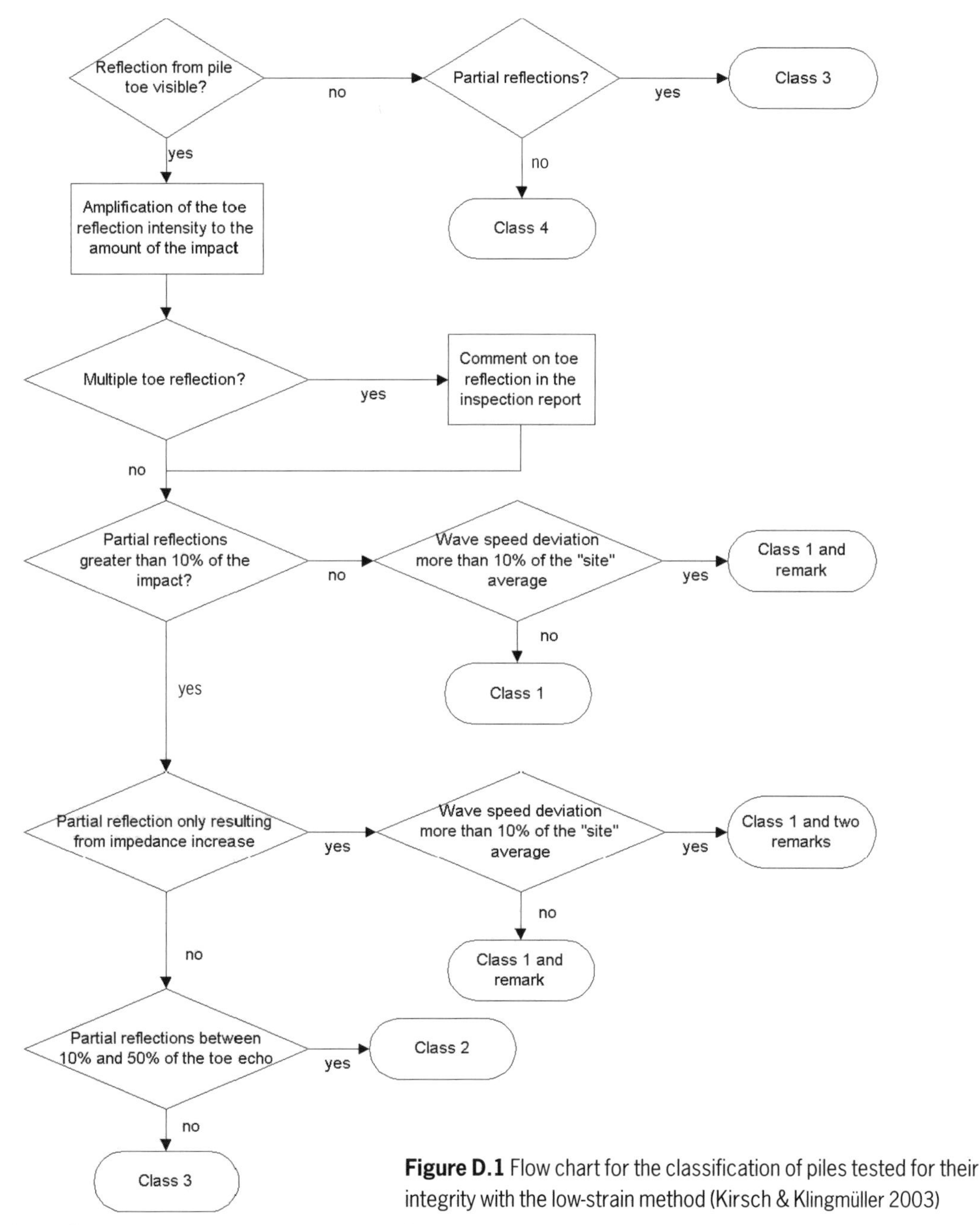

Figure D.1 Flow chart for the classification of piles tested for their integrity with the low-strain method (Kirsch & Klingmüller 2003)

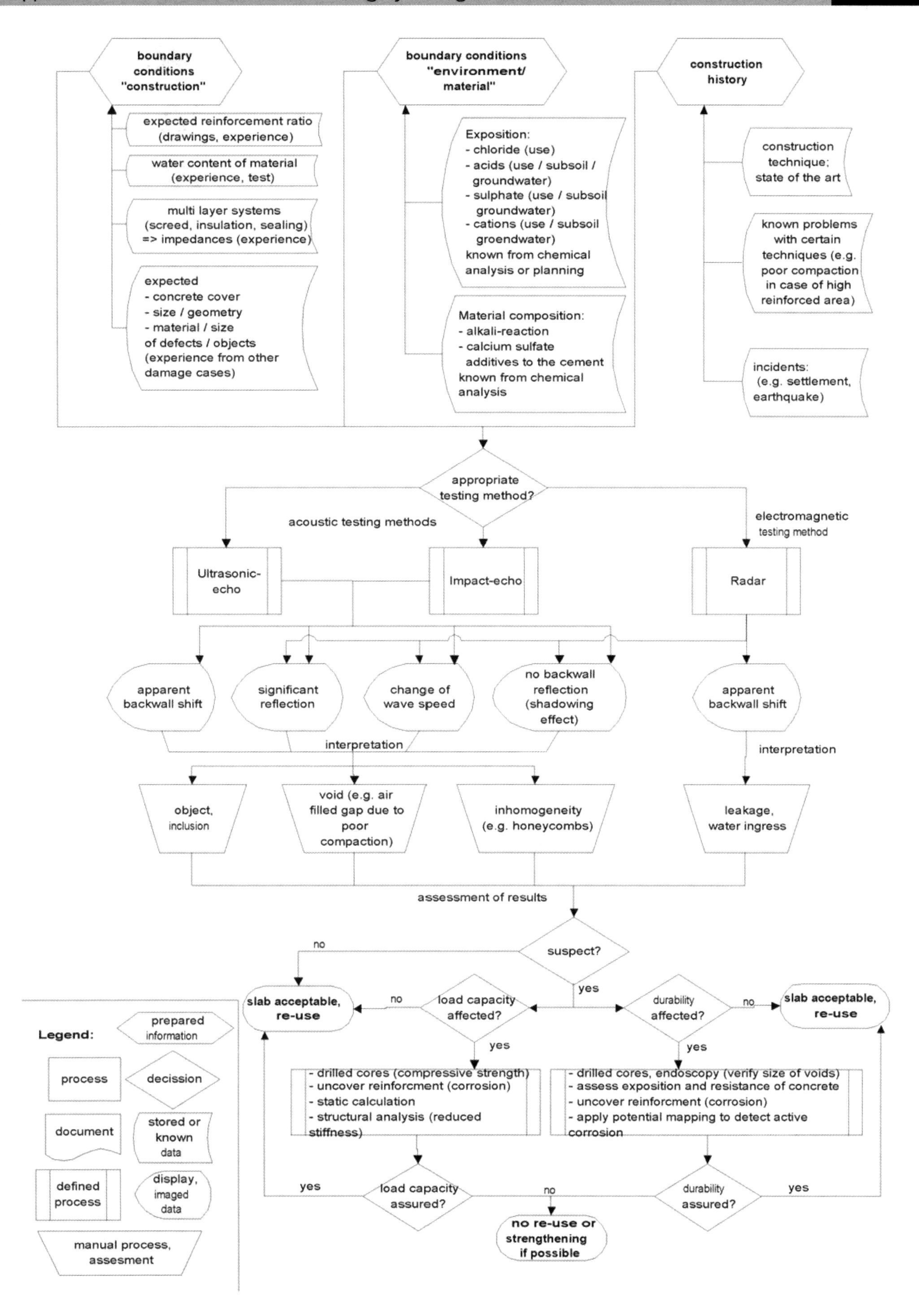

Figure D.2 Testing protocol for integrity testing of slabs with ultrasonic echo, impact echo and radar

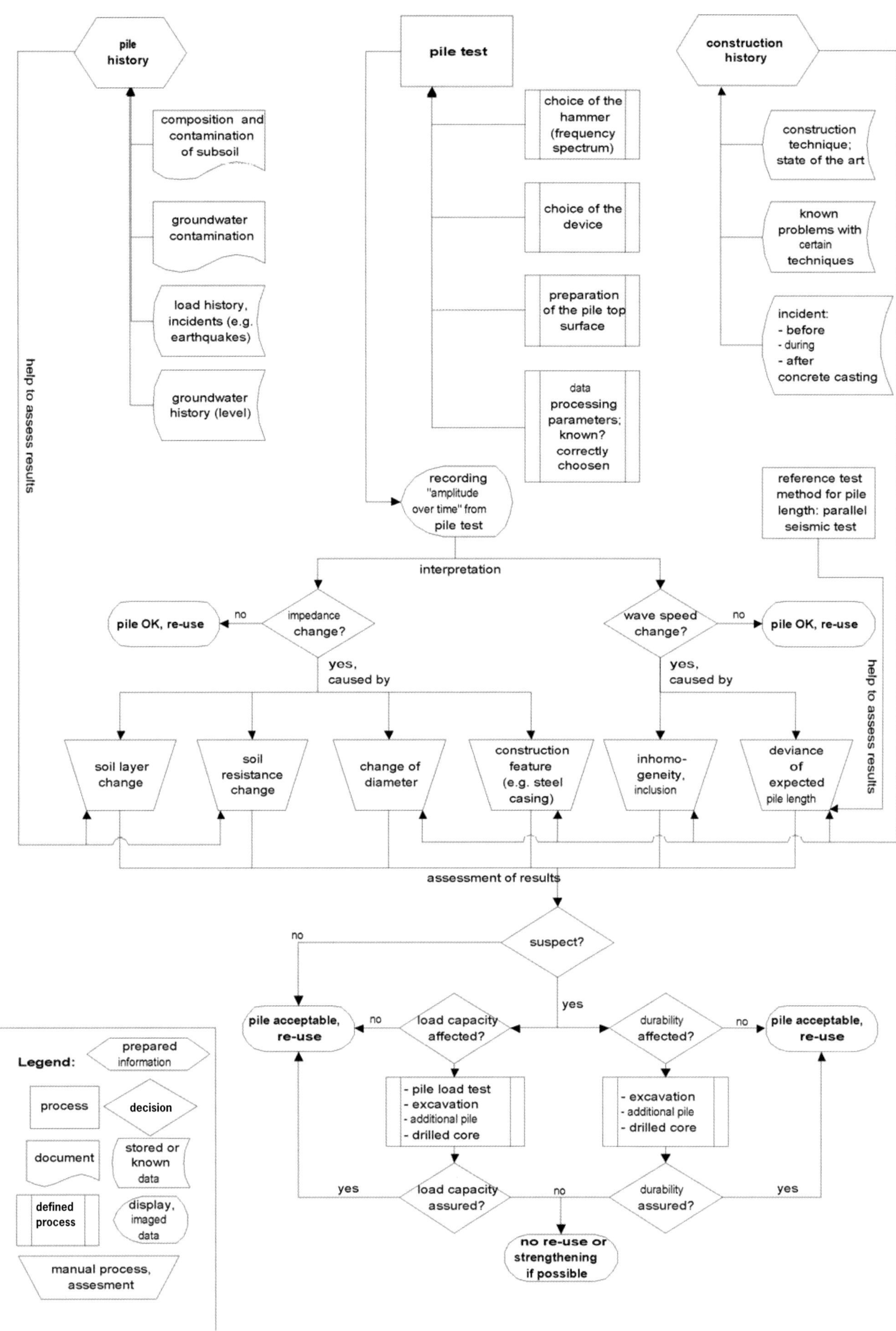

Figure D.3 Testing protocol for integrity testing of pile foundations

Table D.1 Guidelines for ultrasonic echo testing

Thickness measurement:

With increasing thickness of the concrete member choose decreasing frequency:

- Longitudinal waves: f < 85 kHz for thickness > 50 cm
- Transverse waves: f=25..55 kHz for thickness > 100 cm

Prefer transverse waves with low frequency but limited resolution of objects like rebars, voids (or tendon ducts)

Location of objects and voids:

With decreasing size of the object to be detected increase the frequency considering that the penetration depth decreases.
Choose frequencies according to: $f > c_{wave} / (2 \cdot d_{min,object})$

- With $c_{wave} \approx$ 4.000..4.500 m/s for longitudinal waves
- With $c_{wave} \approx$ 2.500..2.800 m/s for transverse waves

Every sensor has a “dead zone” in the surface near areas, where no objects can be detected. Depending on the device this zone can vary from 5 to 10 cm depth.

Positioning of the sensors:

Locate the upper reinforcement layer with a standard testing method
Position sensors in the space between the rebars.
No extra layer should be on the concrete surface. If there is a screed with good acoustic contact quality of results will decrease. If debonded or containing air (e.g. sealing layer) no measurement will be possible.

Dry point contact sensors for transverse waves:	**Array of plane sensors** for longitudinal waves:
▪ No need for coupling agent ▪ Unevenness up to ±5 mm ▪ Suitable automated testing ▪ To improve signal/noise ratio many positions and reconstruction calculation necessary	▪ Coupling agent needed ▪ Unevenness up to ±1 mm ▪ Good signal/noise ratio in a single measurement due to various transmitter/receiver positions within the array

Table D.2 Guidelines for impact echo testing

Interrelation impact / testing frequencies:

The useful frequencies of a spectrum from an impacting steel sphere with D_{sphere} can be calculated according to empiric relation (Hasenstab et al, 2004): f_{max} [Hz] = 291 / D_{sphere} [m]

For practical testing with $c_{impact\text{-}echo} \cong 4.000$ m/s:

- Sphere ∅3 mm: useful frequency up to ≃100 kHz => minimum wavelength λ_{min}=40 mm
  Measurement possible in depth greater $d_{min} > \lambda_{min}/2 = 20$ mm
- Sphere ∅15 mm: useful frequency up to ≃20 kHz => minimum wavelength λ_{min}=200 mm
  Measurement possible in depth greater $d_{min} > \lambda_{min}/2 = 100$ mm

The appropriate choice of diameter D_{sphere} of the impactor is the most important parameter for t he depth range that can be reached by a measurement.

Thickness measurement:

- Thickness measurement of thin concrete members: small sphere diameter
- Thickness measurement of thicker concrete members: larger sphere diameter
- Compared with the true longitudinal wave speed c the apparent wave speed $c_{impact\text{-}echo}$ has to be corrected with a shape factor β regarding P-wave thickness mode of vibration. For a solid plate β is given in Hasenstab et al (2004) with 0.96.

$$d = \frac{c_{impact-echo}}{2f} = \frac{0.96 \cdot c_{long}}{2f} \quad \text{with} \quad c_{long} = \sqrt{\frac{E_{dyn} \cdot (1-\mu)}{\rho \cdot (1+\mu) \cdot (1-2\mu)}}$$

Location of objects and voids:

- Location of surface near objects: small sphere diameter
- Location of objects in greater depth: larger sphere diameter considering limited resolution with decreasing frequency (increasing wavelength)
- With decreasing size of the object to be detected increase the frequency/decrease wave length (smaller sphere diameter) considering that the penetration depth decreases.

For a successful detection of objects that produce a significant reflection at their boundary their spatial extent e should be bigger than their depth d (e/d >1). Voids in concrete members, e.g. honeycombing, often show great variation.

Positioning of the sensors:

- No coupling agent is needed
- On rough surfaces just the accelerometer needs good acoustic contact
- No extra layer should be on the concrete surface. If there is a screed with good acoustic contact quality of results will decrease. If debonded or containing air (e.g. sealing layer) no measurement will be possible.
- With hand-held impactors many measurements can be carried out in a short time.

Table D.3 Guidelines for parallel-seismic measurements

Norms, Regulations, Recommendations:

Only France has set up a norm (NF P 94-160-3). Recommendations are available in UK (Turner, 1997) and USA (FHWA, 2003).

Pile length measurement:

Application range: 5 to more than 50 m

Uncertainty: depends on distance borehole/pile and interpretation method. Distance less than 1 m: 2%. Distance is less than 3 m: Less than 10% (classical interpretation method), less than 3% (improved interpretation methods).

Requirements

a) Source

Impact soure (e. g. sledge hammer), impact time less than 1 ms.

b) Sensors

Hydrophones or borehole geophones, frequency range 10 Hz – 5 kHz. Depth interval 0,3 to 1 m.

c) Recording equipment

Sampling interval < 0.05 ms (20 kHz). Relative trigger accuracy (if measurements done in several steps) better than 0.05 ms.

d) borehole

Distance to pile less than 3 m. PVC or PE casing. Space between casing and soil has to be grouted to provide acoustic contact. Best method: fill borehole half with concrete before inserting casing (with cap at bottom). If using hydrophones as sensors, casing has to be filled with water. Casing diameter depends on sensors (recommendation: sensor diameter + 1-2 cm).

Table D.4 Guidelines for low-strain testing

Interrelation impact (hammer blow) / testing frequencies:

The optimum impact time is limited by the geometry of the pile (diameter d; pile length l):

$d_{pile} \leq \lambda_{limit} / 2 \leq L_{pile}$ with $\lambda_{limit} = t_{impact} * c_{dilat}$
t_{impact} = impact time; c_{dilat} = dilatational wave speed

$$c_{dilat} = \sqrt{\frac{E_{dyn}}{\rho}}$$

$c_{dilat} \cong$ 3.500..4.200 m/s for concrete piles

Typical impact times: Elway (1997)

- steel sphere ∅10..20 mm (0,07..0,15 ms);
- hammer with steel cap (0,50 ms);
- hammer with plastic cap (0,84..1,01 ms)

- Hammer blow with a hard cap not too short because disturbances due to transverse vibrations
- Hard hammer cap => short impact time => good resolution of smaller voids but limited penetration depth
- Soft hammer cap => long impact time => only resolution of larger voids but good penetration depth

---

Pile length measurement:

Application range: pile length 5..20 m; pile length / pile diameter < 30:1

The wave speed in the pile is influenced from the pile shape: screw shape pile show a 8-12%decreased wave speed compared with the expected c_{dilat} of a straight pile shaft. Piles with a metal liner at the top show an 6-15% increased wave speed.

Detectable defects*:
From large-scale model tests on reinforced concrete piles presented in Elway (1997) a summary is given:

- **Minimum defect length for a loss of section** from 60% (further called defect): $l_{defect} \geq \lambda_{limit} / 40$ (for t_{impact} = 0,4 ms, f_{limit} = 2,5 kHz, c_d = 3.500 m/s: l_{defect} = 3,5 cm)
- **Critical defect length**: $l_{defect} \geq \lambda_{limit} / 4$ (for t_{impact} = 0,4 ms, f_{limit} = 2,5 kHz, c_p = 3.500 m/s: l_{defect} = 35 cm). At smaller defect lengths or lower frequency contents the exact determination of the defect length is not possible.
- Differences between **symmetrical or unsymmetrical loss of section** could not be detected.
- In case of **continuously loss of section** the length of the defect has to be interpreted as an average value of the total length.
- Relating to the $\lambda_{limit}/4$-criteria **several defects** along the pile are theoretically distinguishable if their distance is more than $\lambda_{limit}/4$. But defects in the upper pile area may cause multiple reflections in greater depth that may hide other true reflections or cause apparent but false reflections and lead to misinterpretation.

*The above given criteria may give a hint if an expected defect might be detected.

Preparation of the pile head:

- Loose concrete at the pile head has to be removed
- Surface has to be dry, preferably grinded at sensor position
- No extra layer should be above
- Force-fit and fixed coupling of the sensor

Table D.5 Guidelines for radar testing

Thickness measurement:

Typical antenna frequencies for thickness measurement or depth position of a reflector:

- 40..100 cm: f = 500 MHz, 900 MHz (decreasing penetration depth with increasing f)
- < 40 cm: f = 1,5 GHz or higher

Avoid upper reinforcement layer with spacing less than 10 cm.
Wet or young concrete allows lower penetration depth.

Location of objects and voids:

The choice of the direction of the electric field (antenna polarization) in relation to the reflector that should be detected allows the best results:

- Detection of a linear reflector (e.g. reinforcement): electric field parallel to reflector
- Detection of a reflector beneath a linear reflector (e.g. back wall beneath reinforcement layer): electric field perpendicular to linear reflector

For testing a certain area it is recommended to measure the area with both polarizations.

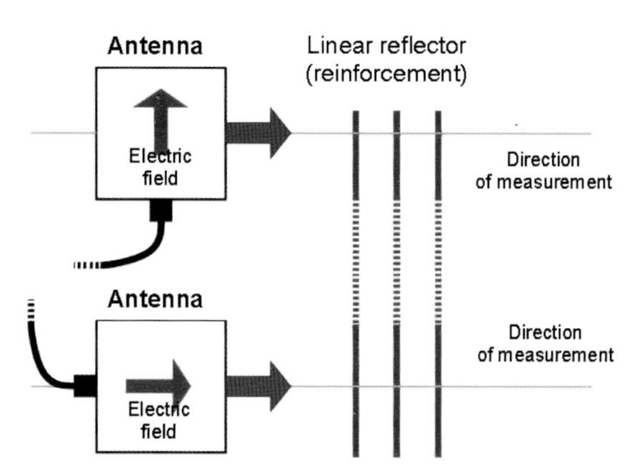

Positioning of the antenna and measuring grid:

- Choose antenna polarization according to previous paragraph
- Surface of tested concrete member should be dry (no water standing on the surface)
- Measuring grid for location of objects or geometry detection: lines with 5..25 cm spacing; perpendicular.
- Measuring grid for thickness measurement: one line representing a representative profile.
- Along every measuring line an A-scan should be triggered every 0,5 mm..2,0 cm depending on the wanted spatial resolution.

D.2 References

Ellway K. Practical guidance on the use of integrity tests for the quality control ofcast-in-situ piles. *Proceedings of International Conference on Foundations and Tunnels*, London, March 1987. pp 228–234

Federal Highway Agency. *Federal Lands Highway Program: Geophysical Methods*. 2003 http://www.cflhd.gov/agm/engApplications/BridgeSystemSubstructure/212BoreholeNondestMethods.htm

Hasenstab A, Krause M, Hillemeier B & Rieck C. Materialuntersuchungen an Holz mit niederfrequenter Ultraschall Echo Technik [Material testing of timber structures with low-frequent ultrasonic-echo technique]. In: *DACH-Jahrestagung*, 17–19 May 2004

Salzburg A. CD-ROM Vortrag 87. Berlin, DGZfP

Kirsch F & Klingmüller O. Erfahrungen aus 25 Jahren Pfahl-Integritätsprüfung in Deutschland [25-years experience with pile-integrity testing in Germany]. *Bautechnik* 2003: ***80*** (9): 640–650

Assosciation francaise de normalisation. Auscultation d'un 'el'ement de fondation. Partie 3: M'ethode sismique parallèle (MSP). NF P 94-160-3, 1993

Turner MJ. *Integrity testing in piling practice.* CIRIA Report 144, 1997

Appendix E: Instrumentation case studies

Case 1: Bankside, London, UK

Responsibility for instrumentation

Cementation Foundations Skanska, UK

Buildings

The Bankside 123 project is the redevelopment of a site previously occupied by a Ministry of Defence building in Southwark, London. The previous structure consisted of an office building with a single-storey basement. The new structure has up to 14 floors and up to 3 storeys of basement.

Foundation

The previous structure was supported on a large number of 450 mm diameter concrete cast-in-situ piles, up to approximately 10 m in length. The existing basement slab was removed. Piles that clashed with the new pile positions were removed. The foundation work consisted of the installation of 750/600 mm diameter secant wall piles and 1200–2400 mm diameter bearing piles.

Objectives of instrumentation

- To investigate the behaviour of the capping beam, secant pile wall, bearing pile and pile cap foundation elements and over-consolidated clay strata during basement excavation, superstructure construction and building life-cycle.
- To trial new fibreoptic strain sensors alongside more traditional vibrating wire strain gauges.

Instrumentation and sensor locations

For structure overview and sensor location, see Figure E.1.

- A section of the capping beam and secant wall was instrumented radiating out beneath a column loading on top of the capping beam. Measurements were taken of the horizontal strain in the capping beam to investigate capping beam load spread using vibrating wire strain gauges across the capping beam section and spaced at 3 m intervals.
- Vibrating wire strain gauges were installed at 3 m spacing along the pile length in 6 piles (instrumentation in 2 piles to 19 m depth, 4 piles to 12 m depth) to investigate reserve load capacity within the secant pile wall. A vibrating wire strain gauge attached to the pile reinforcement cage is shown in Figure E.2.
- Pressure cells were installed between the pile cap and the soil underneath a ground bearing pile cap to investigate the interaction between pile and cap and investigate potential

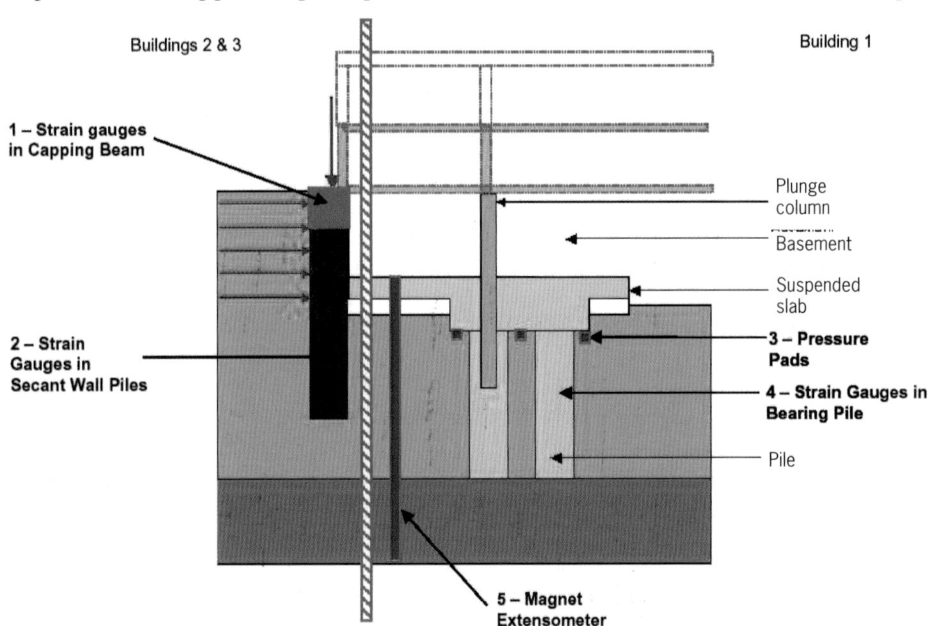

Figure E.1 Overview of structure and sensor placements

Figure E.2 An attached vibrating wire strain gauge

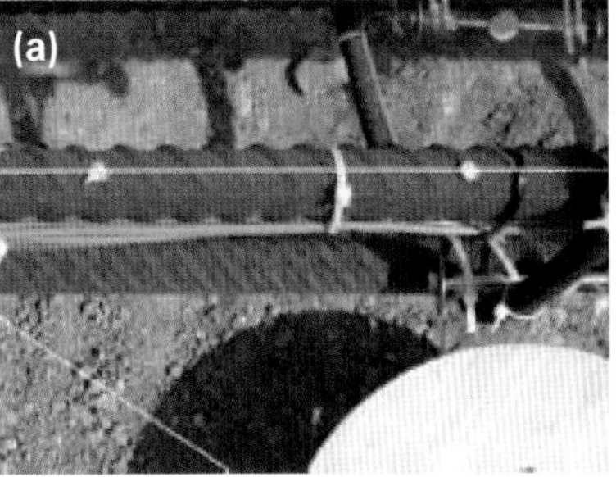

Figure E.3 (a) Cambridge University Brillouin system; **(b)** City/Cranfield University Bragg grating system

reserve capacity. The pile cap chosen also included an instrumented bearing pile.

- In collaboration with Cambridge University and City and Cranfield Universities, two bearing piles were instrumented with trial fibreoptic instrumentation. The Cambridge fibreoptic instrumentation used the Brillouin (BOTDR) technique to measure distributed strain along the cable. The cables were installed attached to the pile reinforcement cage in pre-tension to allow measurement of both tensile and compressive strains. The City/Cranfield fibreoptic instrumentation used Bragg grating discrete sensors, pre-attached to steel bars which were inserted into the pile reinforcement cage. These sensors also measured vertical strain in the piles. The fibreoptic instrumentation is shown in Figure E.3. To compare the instrumentation with traditional vibrating wire strain gauge instruments, levels of four strain gauges at 3 m centres were installed along the full pile length. Each had a single cable which was brought to the pile surface (as for the fibreoptics) and these have been connected via intermediate loggers to the main logger for remote interrogation via mobile modem connection and off-site data storage.
- To 'complete the loop' and ensure that there is a context within which to assess pile strain, a magnet extensometer was installed to measure ground movements during the excavation (unloading) and construction (re-loading) phases. Base readings have been taken to allow comparison with later stages.

Results

Figure E.4(a) shows the changes in strain from a datum date, recorded by the three different strain sensor types at three discrete levels in a pile. It appears that there is a good correlation between the three sensor types except for a shift in strain that occurs for the Bragg grating sensors in February 2005 and noticeable 'end effects' at the pile toe. Figure E.4(b) illustrates the power of a distributed sensor in geotechnical applications and shows how the discrete sensors can produce accurate local strain measurements and yet fail to illustrate adequately the pertinent strain profile, whereas the distributed method allows the user to gain a fuller picture of the pile reaction to loading. Results from the monitoring at the Bankside 123 development are included in Tester et al (2006).

Comments

- Instrumentation enabled the provision of an excellent case study for soil–structure interaction to allow development of understanding of potential reserve capacity for over-consolidated clay sites.
- The fibreoptic systems were installed successfully, although the importance of the whole project team working together and planning the installation and subsequent sensor/cable protection must be emphasised.
- Reasonable correlation between the two fibreoptic systems and vibrating wire strain gauges was observed.
- Specialists will be required to analyse the data produced and assess unusual features in the data, along with the potential causes. This is not a simple instrumentation situation.

(a)

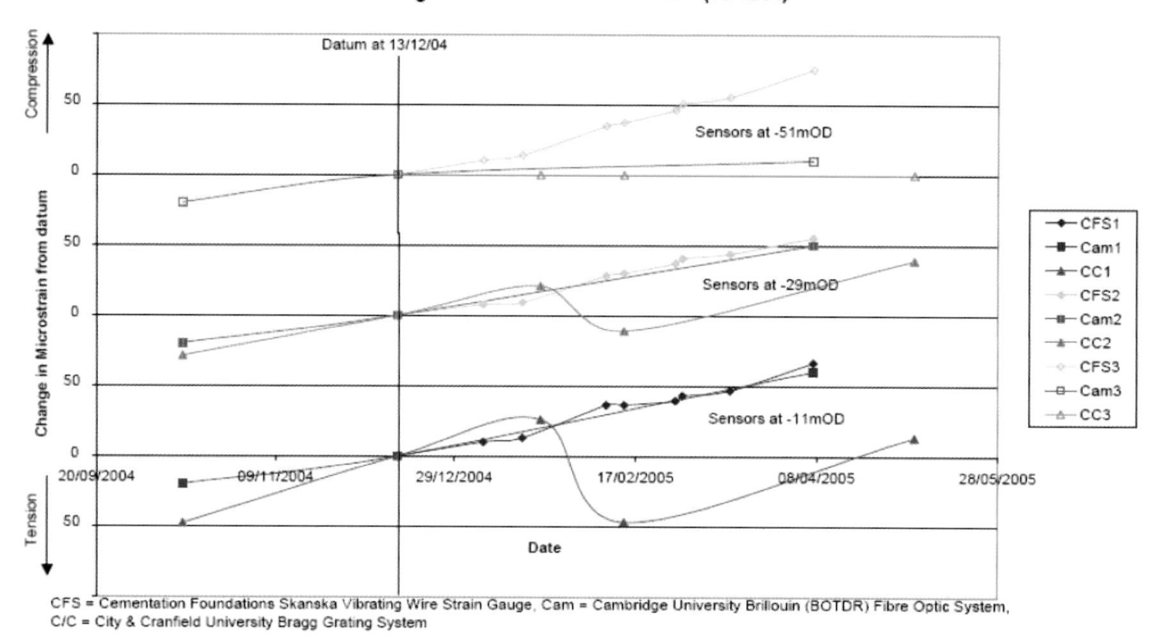

(b) **Pile 11 strain sensor data**

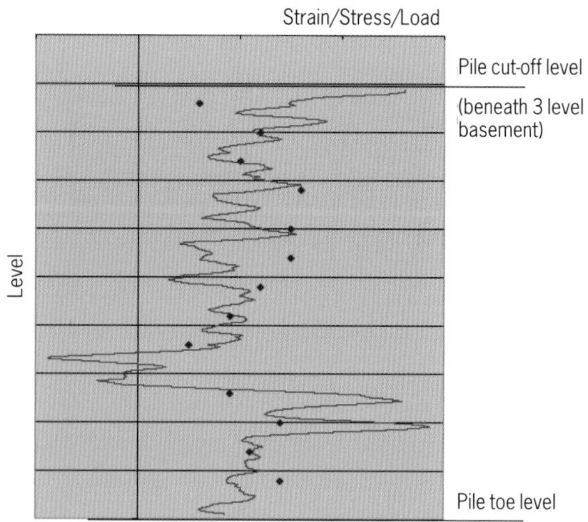

Figure E.4 (a) Comparison of changes in strain recorded by discrete and distributed fibreoptic strain sensors and vibrating wire strain gauges; **(b)** comparison of Brillouin and vibrating wire strain sensor readings from a pile at Bankside 123

Case 2: Zlote Tarasy, Warsaw, Poland

Figure E.5 Zlote Tarasy: final project view

Responsible for instrumentation

Soletanche Bachy, France

Building

A 23-storey tower

Foundation

Combined pile-raft foundation. 900 mm diameter cast-in-situ piles and D-Walls

Objectives of instrumentation

Evaluate the actual loads in the piles and the raft during the construction of the tower.

Instrumentation

- To evaluate the load distribution, 7 pile load cells (one per instrumented pile) and 7 raft load cells were installed. The electrical pile load cells consisted of a ring-shaped stainless steel body fitted with strain gauges. These cells, with their load distribution plate, were installed at the top of the pile.
- To study the load distribution in the raft, 15 vibrating wire strain gauges were installed on the raft reinforcement rebars close to the pile and raft load cells. Vibrating wire strain gauges (Figure E.6), especially designed for long-term strain measurements in mass concrete were used. However, the planned installation was strongly disturbed and reduced by the site operations and the raft loads could not be monitored exactly as initially planned. Accordingly, the strain gauge measurements were not used for the study.
- Automated data collection system linked to the Internet was established.

The measurements were combined together to study the stress evolutions in the raft and piles, respectively, during construction of the tower. The locations of the sensors are shown in Figure E.7. The installation of a strain gauge in the pile head is shown in Figure E.8.

Results

For the pile load cells and for the raft load cells, the measurements show as an increase in load since the beginning of loading. In Figure E.9 the slab level, pile loads and raft loads are presented versus time. The evolution of the tower construction is gradually increasing the loading on each pile. The loads in the raft are increasing for the first part of the loading and then showing a small increase, rather than staying constant as expected.

A load transfer from the raft (load taken initially by the soil) to the pile could be observed over a few days after the pouring of the concrete.

All recorded loads were smaller than initially anticipated.

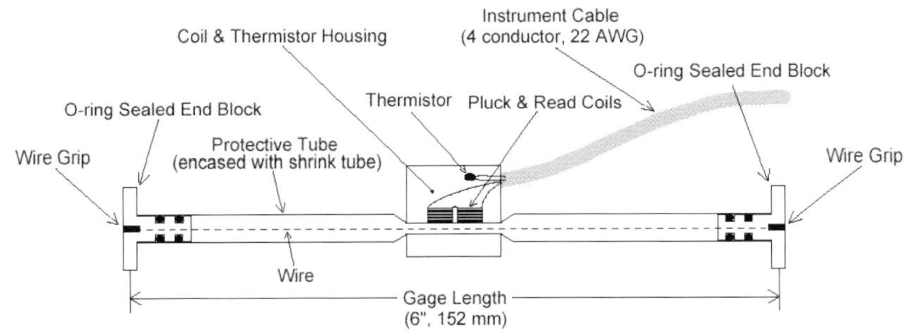

Figure E.6 Model VCE-4200 Vibrating wire strain gauge

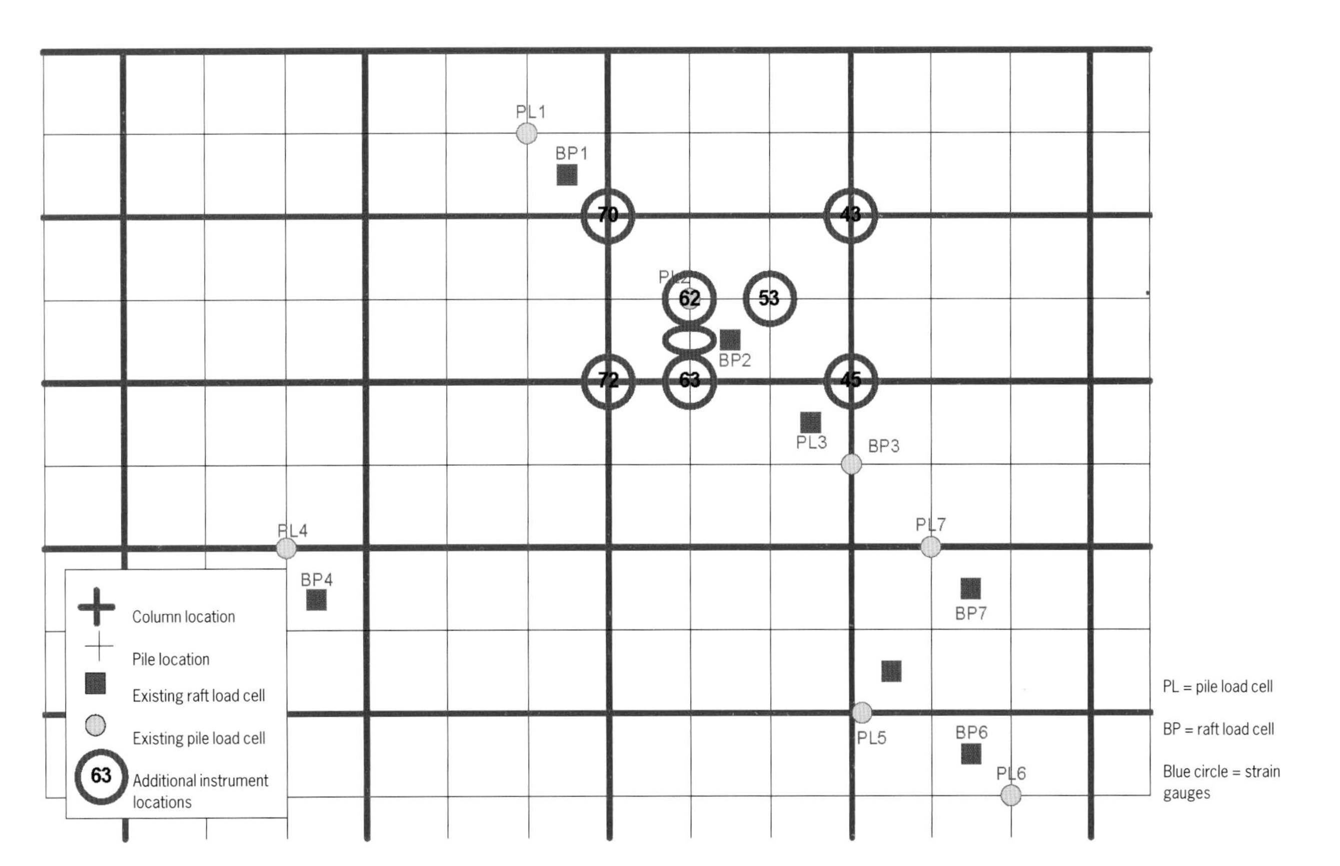

Figure E.7 Sensor locations

Figure E.8 Strain gauge in pile head

Comments

- It was unfortunate that the construction sequence disturbed the installation and monitoring process and that strains could not be recorded exactly as initially anticipated. The fact that the installation was part of a research project and was not vital for the site staff, was an additional problem. However, such problems also sometimes arise when a site requires monitoring. When planning instrumentation, it is important to consider redundancy .
- It is important (if possible) to consider possible modifications to the construction scheme.
- The required sensor placement and protection must always be carefully considered.
- The load cell measurements, both for the pile and raft loads, appear to be reliable and useful.
- When installing raft load cell it is important that the soil below the cell is representative for the site.

Reference

Tester PD, Fernie R, Bennett PJ, Kister G & Gebremichael Y. Brillouin and Fibre Bragg Grating fibre-optics development at a RuFUS site in London. In: Butcher AP, Powell JJM & Skinner HD (eds). *Reuse of Foundations for Urban Sites: Proceedings of International Conference*, BRE, Watford, 19–20 October 2006. Bracknell, IHS BRE Press, 2006

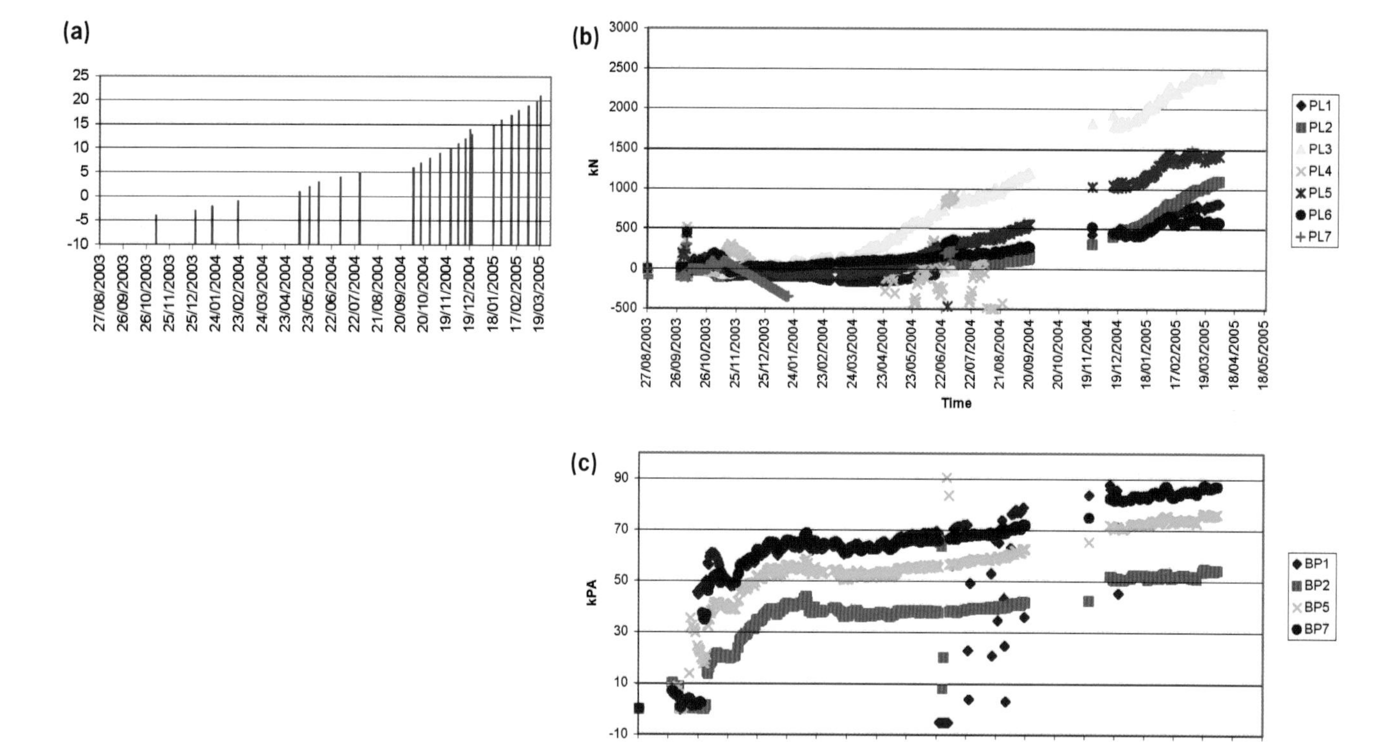

Figure E.9 Results of monitoring: **(a)** slab level, **(b)** pile load cells and **(c)** raft load cells, with time

Appendix F: Example documentation from a typical pile foundation project

F.1 Introduction

After the completion of the foundation works, a close-out report is generally prepared. All factual information and any relevant interpretative and descriptive information associated with the piling works should be gathered into the close-out report and all the necessary documents included (eg pile construction details, pile record sheet, test records, non-conformance report, relevant e-mails, faxes, etc.).

Below follows extracted parts (highlights) of a piling supervision close-out report for an actual pile foundation system for a building.

The extracted report below can be adapted to the structure described in *Chapter 7* although this example includes only the key elements that can be found in a close-out report and adapted to metadata.

Figures are not included in this example as its purpose is to illustrate typical coverage.

F.2 Piling contract

F.2.1 Project organisation

The project organisational structure for the piling package included the following personnel:

- Principal contractor,
- Planning supervisor,
- Consulting engineer,
- Piling contractor,
- Structural engineer.

F.2.2 Schedule of works

The actual start and completion dates of the three main activities associated with the pile construction are shown in Table F.1.

Table F.1 Construction dates

Activity	Actual start date	Actual finish date
Piling		
Low level	25 Sep 2000	12 Jan 2001
High level	25 Jan 2001	16 Mar 2001
Base grouting		
Low level	12 Oct 2000	3 Apr 2001
High level	12 Mar 2001	28 Mar 2001
Sonic logging		
Low level	4 Oct 2000	4 April 2001
High level	20 Feb 2001	21 Mar 2001

F.3 Site

F.3.1 Design stratigraphy

The stratigraphy shown in Table F.2 was used for the design of the pile foundations. Full descriptions of the soil units would be included in the Geotechnical Report.

Table F.2 Design stratigraphy

Stratum	Reduced level of top of stratum (mOD)	Thickness of of stratum (m)
Made ground (high level only)	+6	12
Terrace Gravel (mainly removed)	–6.0	3.0
Lambeth Clay	–9.0	4.5
Lower Lambeth Sand	–13.5	5.0
Thanet Sand	–18.5	15.0
Chalk	–33.5	Not proven

F.3.2 As- built stratigraphy

Strata levels were checked by the pile contractor's banksmen and occasionally checked by the resident engineer. Generally, the design stratigraphy was confirmed. Any deviations were recorded.

F.3.3 Groundwater

The groundwater regime beneath the site consisted of an upper and lower aquifer.

F.4 Pile design

F.4.1 Specification and drawings

The structural engineer produced the pile layout and typical pile detail drawings listed as follows.

Table F.3 Pile design summary

Pile type	Diameter (mm)	Cut-off level (mOD)	Toe level (mOD)	Minimum Thanet Sand penetration (m)
BP1	900	+3.6 or + 4.1	-22.5	4
BP2	1050	+3.6	-22.5	4
BP3	1200	-5.5	-26.5	8
BP5 or BP6	1500	-6.5	-26.5	8

- 440/00001 – SP1 DS4 Pile layout
- 440/00002 – SP2 DS4 Typical pile details

Note: The final pile layout and typical pile details should be recorded in the close-out report.

A request from the pile contractor was submitted to the consulting engineer to reduce the pile diameter from 1050 mm to 900 mm on as many high level piles as possible.

F.4.2 Design issues

Originally, the pile design philosophy was based on the results of the preliminary test pile at D (Feb 1999) and pile test data. A summary of the pile toe depths and sand penetrations is included in Table F.3.

Subsequent to the performance of the contract pile load tests, the pile design basis was reviewed and revised. The new pile capacities were detailed in the revised calculations and were recorded in the table on the pile details drawing in the close-out report.

F.5 Pile installation

F.5.1 Contractors method statement

The D piling works were completed in accordance with the method statements for D listed below.

- Construction issue method statement
- Addendum method statement: Pile position probing
- Addendum method statement: Construction of high-level piles using vibrated temporary casings
- Base grouting method statement
- Method statement: Rigging/de-rigging of pile test reaction assembly
- Method statement for static pile load testing using electronic data logging

F.5.2 Construction summary

Table F.4 summarises the pile construction details. Full details relating to the construction of individual piles should be included in the close-out report.

F.5.3 Influence of groundwater

The stratigraphy of the site and the presence of the perched water required the installation of a temporary casing for each pile. An 8 m long temporary casing was used for this purpose and provided an adequate seal for all piles. Both continuous and segmental casing was used on at high level to provide support to the made ground.

F.5.4 Construction procedures

Low level piles (P20–P177)

A temporary casing was installed (see *Section F.5.3)*, thus protecting the pile bore from intrusion of water from the upper aquifer. Coupled with the lowered water table of the deep aquifer, this allowed the piles to be drilled to full depth using an auger without the use of bentonite. The final 200–300 mm of each pile was excavated with a cleaning bucket to ensure an even, clean base.

High level piles (P1–P19)

Casing installation and drilling of the pile bore were undertaken as above.

Table F.4 Summary of piles constructed

Description	Number	Comments
Contract piles tender for	171	—
Contract piles omitted post tender	10	Not required because of changes in building design
Contract piles added post tender	10	Required because of changes in building design
Contract pile type changed	8	Changed from 1050 mmϕ to 900 mmϕ
Replacement piles constructed	1	P2a
Contract piles downgraded	3	P2, P11, P12
Anchor piles constructed	9	Temporary piles for contract pile testing
Piles base grouted	172	—
Piles sonic logged	172	—
Piles re-sonic logged	11	—
Non-conformance reports (NCRs) issued	13	—

F.6 Quality

F.6.1 Quality management

The quality system used should be quoted in the close-out report.

F.6.2 Contractor's quality control

Reference should be made to the pile contractor's *Quality plan: construction issue quality plan.*

The following record sheets were completed for each pile by the pile contractor:

- bored pile record sheet,
- concrete record sheet,
- reinforcement cage record,
- bentonite testing record (if required),
- procedure check sheet,
- pile bore concrete volumes.

Base grouting records were also completed for each pile and contained the following information:

- base grouting information sheet,
- transducer uplift plots,
- pressuremeter and flowmeter grout module instrumentation plots.

A sample of the pile record sheets submitted by the pile contractor should be included, with the rest available in the close-out report.

F.6.3 The consulting engineers quality assurance

Site supervision strategy

The supervision strategy employed at site was to carry out systematic checks of the contractor's work, to monitor the contractor's quality systems and to focus on any problems as they occurred. The consulting engineer monitored the following activities during the piling works:

- casing seal and verticality,
- Thanet Sand level during boring,
- measurement of the pile depth at key stages of construction,
- bentonite sand content testing,
- reinforcing cage construction and installation,
- concreting,
- base grouting,
- sonic logging,
- contract pile testing.

In addition, regular checks were undertaken to ensure that the boring tools were within the required tolerance. All results were recorded.

F.6.4 Materials testing

The following materials were periodically checked and records kept to ensure compliance with the specification and the pile contractors method statement and quality plan:

- bentonite,
- grout,
- reinforcing steel,
- concrete (slump tests and cube strengths).

F.6.5 Contractor's records

The following records were submitted by the pile contractor for the consulting engineer's comments:

- piling records,
- sonic logging records,
- base grouting records,
- concrete cube records.

The consulting engineer on behalf of client signed all of the pile contractor's records for record purposes only.

F.6.7 Significant quality issues

A summary of the non-conformance reports (NCRs) issued by the pile contractor should be included in the close-out report.

Copies of all NCRs, as submitted by the pile contractor, were included. A total of 13 NCRs were issued. On receipt of an NCR from the principal contractor, the pile contractor returned the NCR form to the principal contractor for comment, proposing remedial actions where necessary. A copy of the NCR, including the proposal action, was issued to the relevant consultant. NCRs relating to pile reinforcement were sent to the structural engineer and other aspects sent to the consulting engineer. NCRs were closed-out following completion of acceptable remedial action or a review of design requirements. The pile record sign-off sheet, lists all piles and any NCRs that may have affected the individual piles.

F.7 Base grouting

F.7.1 Summary

All of the contract piles and one replacement pile were base-grouted. Base grouting was carried out using four independent circuits of the tube à manchette (TAM) type. A summary of all base grouting records is included.

F.7.2 Contactor's requirements

The contractor worked to the Base grouting method statement included.

The contractor submitted a summary of the base grouting performance and grout module data for each pile.

F.7.3 Uplift and volume criteria

The Particular Specification for Piles stated the following base grouting criteria.

- At the completion of base grouting, the pile uplift shall not be less than 0.2 mm and shall not exceed 2.0 mm. If the minimum uplift (0.2 mm) criterion is not met on the first grouting, the contractor shall re-grout the pile.
- The following pressure-grouting criteria shall be satisfied:
  1. a minimum pressure of 30 bars on two opposite grouting circuits. The residual grout pressure of 15 bars to be held at least 2 minutes on two opposite grouting circuits.

2 a minimum grout volume of 25 litres/circuit.

If criteria 1 and 2 are not met during the first or subsequent grouting phases the pile shall be re-evaluated.

F.7.4 Possible causes of low uplift

A report of deviations is included and a short possible explanation to the deviations is given.

F.7.5 Investigations of piles with low uplift

The specification calls for *'observable' uplift (0.1 mm)* to be achieved. The pile contractor's two independent monitoring systems were both capable of showing clearly when at least 0.05 mm of uplift had occurred; it is thought that the systems were accurate to a tolerance of at least 0.01 mm. The detailed grouting records for all piles that had less than 0.1 mm uplift were reviewed and no evidence of hydrofracture was observed.

F.7.6 Acceptances

8% of piles on the site contract did not meet the specified base grouting criteria.

Contract load tests were carried out on two of these piles (P060 and P038) and were found to perform satisfactorily.

F.8 Sonic logging

F.8.1 Introduction

Cross-hole sonic logging was carried out to verify the integrity of each pile between the pile base and cut-off level. For details of the test method employed, refer to *CIRIA Report 144* (Turner 1997). All piles were installed with four base grouting circuits. These tubes were also used for the sonic logging. The tube configuration provided four profiles around the edge of the pile shaft and two across the centre.

F.8.2 Evaluation of sonic logging results

The evaluation criteria for sonic logging were based on the recommendations of *CIRIA Report 144* and experience gained on previous projects.

A review of the sonic logs for the site showed major problems on only one pile. However, 11 piles were re-logged for reasons such as evidence of young or weak concrete or loss of signal below the cut-off. All piles that were re-logged showed sound and homogeneous concrete throughout.

F.9 Contract pile testing

F.9.1 Introduction

The purpose of a contract pile test is to show that the load settlement behaviour of the contract pile is similar to that of the preliminary pile tested at DS2 and to act as a check on workmanship and materials.

The Particular Specification required that 1% of the contract piles be subject to a load test. For the D site project, two tests satisfied this requirement. Tests were performed on P90 and P139. The tests were undertaken in accordance with Clause 10.13 of the *Specification for piling and embedded retaining walls* (ICE 1996).

F.9.2 Test loading

As required by *Specification for piling*, contract piles that are load-tested should be loaded to:

- 100% DVL, and then
- 100%DVL + 50% SWL,

where DVL = design verification load and SWL = specified working load.

Due to de-watering of the lower aquifer, the capacity of the pile at the time of testing was up to 30% higher than it would be in the long term. Therefore, to ensure that the load settlement behaviour exhibited by the pile during testing was representative of the pile during its working life, the test load was increased accordingly. Hence, the SWL was taken to equal the DVL and each of the contract piles were loaded to:

- 100% DVL and then,
- 150% DVL.

For each of the pile tests, the DVLs were calculated by the consulting engineer design team and were summarised in the test reports.

F.9.3 Test results

This report includes pile load test reports. Results are summarised, although they should only be read in conjunction with the full discussion in the engineer's report.

F.10 Completion of work

F.10.1 As-built drawings

These should be included in the close-out report.

F.11 References

Institution of Civil Engineers (ICE). *Specification for piling and embedded retaining walls.* London, Thomas Telford, 1996

Turner MJ. *Integrity testing in piling practice.* R144. London, CIRIA, 1997